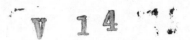

Electrical Blueprint Reading

by

John E. Traister

Craftsman Book Company, 6058 Corte Del Cedro, Carlsbad, CA 92008

Contents

Electrical Drawings

An electrical blueprint is an exact copy or reproduction of an original drawing, consisting of lines, symbols, dimensions, and notations to accurately convey an engineer's design to workmen who install the electrical system on the job. The student should keep in mind that the workmen must be able to take a blueprint, and without further instructions, install or produce the electrical system as the engineer or draftsman intended it to be accomplished. A blueprint, therefore, is an abbreviated language for conveying a large amount of exact, detailed information, which would otherwise take many pages of manuscript or hours of verbal instruction to convey.

In every branch of electrical work, there is often occasion to read an electrical drawing. Electricians, for example, who are responsible for installing the electrical system in a new building, usually consult an electrical drawing to locate the various outlets, the routing of circuits, the location and size of panelboards, and other similar electrical details, in preparing a bid. The electrical estimator of a contracting firm must refer to electrical drawings in order to determine the quantity of material needed. Electricians in industrial plants consult schematic diagrams when wiring electrical controls for machinery. Plant maintenance men use electrical blueprints in troubleshooting. Circuits may be tested and checked against the original drawings to help locate any faulty points in the installation.

TYPES OF ELECTRICAL DRAWINGS

Electrical Construction Drawings

Drawings that represent the physical arrangement and views of specific electrical equipment are called electrical construction drawings. These drawings give all the plan views, elevation views, and other details necessary to construct the job. Fig. 1-1 shows a sketch

of an electrical panelboard "can," or housing. One side of the housing is labeled "front" and another side, "top."

In Fig. 1-2, the drawing labeled "top" is what you see when you look directly down at the top of the housing, and in doing so, the sides, the bottom, the front, and the back are cut from your view.

The drawing labeled "front" is what you see when the block is directly in front of you. In this case, you cannot see the top, bottom, back, or the two sides.

A "side" view is what would be seen if the right side of the housing was turned toward you. This cuts from your view the top, the bottom, the front, the back, and the left sides.

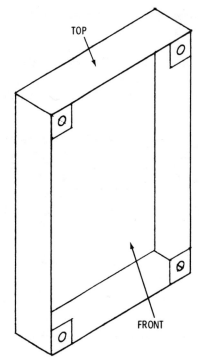

Fig. 1-1. Pictorial view of an electrical panelboard housing.

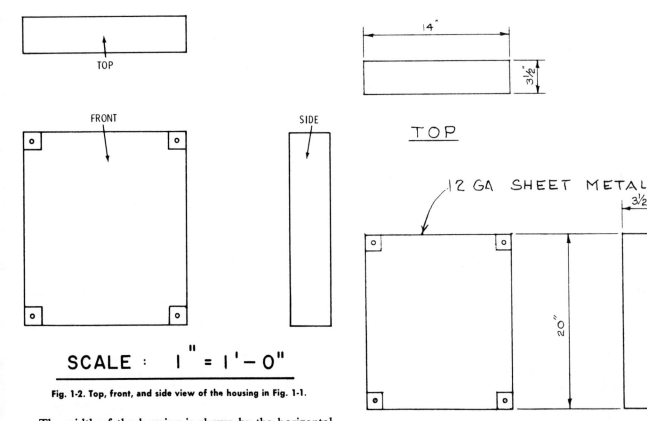

Fig. 1-2. Top, front, and side view of the housing in Fig. 1-1.

Fig. 1-3. Same drawing as Fig. 1-2 with alternate method of showing dimensions.

The width of the housing is shown by the horizontal lines of the top view and the horizontal lines of the front view. The height is shown by the vertical lines of both the front and the side views, and the depth is shown by the vertical lines of the top view and the horizontal lines of the side view.

These three drawings—the top, front, and side views—tell all about the shape of the housing, but they do not indicate the size of the housing. There are two common methods to indicate the actual length, width, and height of the housing. One is to draw these views to some given scale, such as $1'' = 1'\text{-}0''$. This means that 1 inch on the drawing represents 1 foot in the actual con-

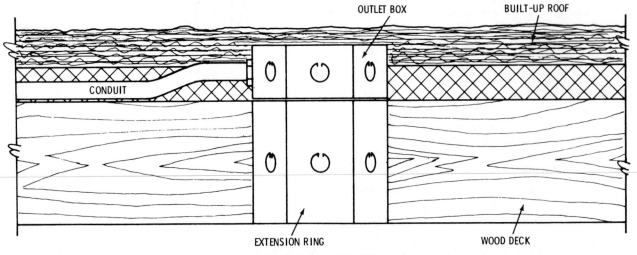

Fig. 1-5. A typical electrical detail drawing.

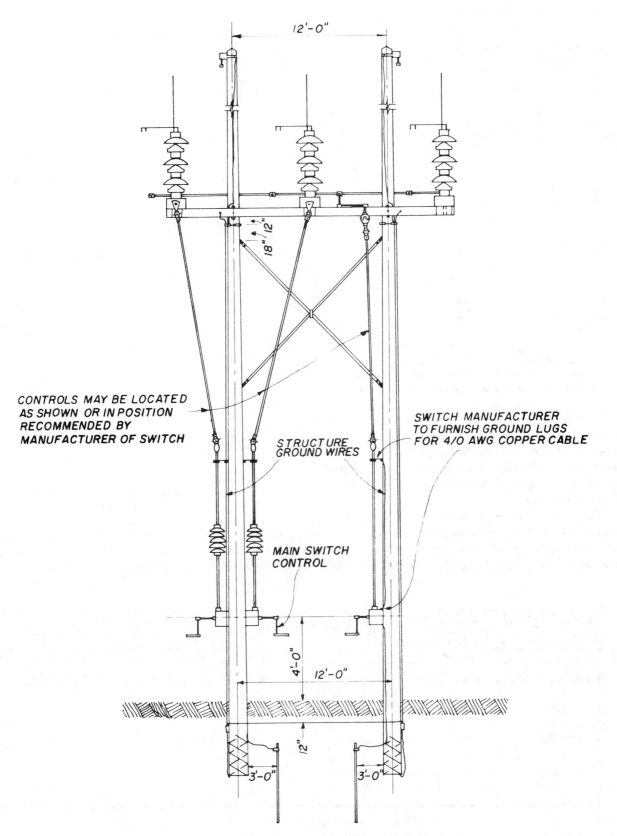

12'-0"

18" 12"

CONTROLS MAY BE LOCATED
AS SHOWN OR IN POSITION
RECOMMENDED BY
MANUFACTURER OF SWITCH

STRUCTURE
GROUND WIRES

SWITCH MANUFACTURER
TO FURNISH GROUND LUGS
FOR 4/0 AWG COPPER CABLE

MAIN SWITCH
CONTROL

4'-0"

12'-0"

12"

3'-0"   3'-0"

Fig. 1-4. Construction detail of a high-voltage transmission line.

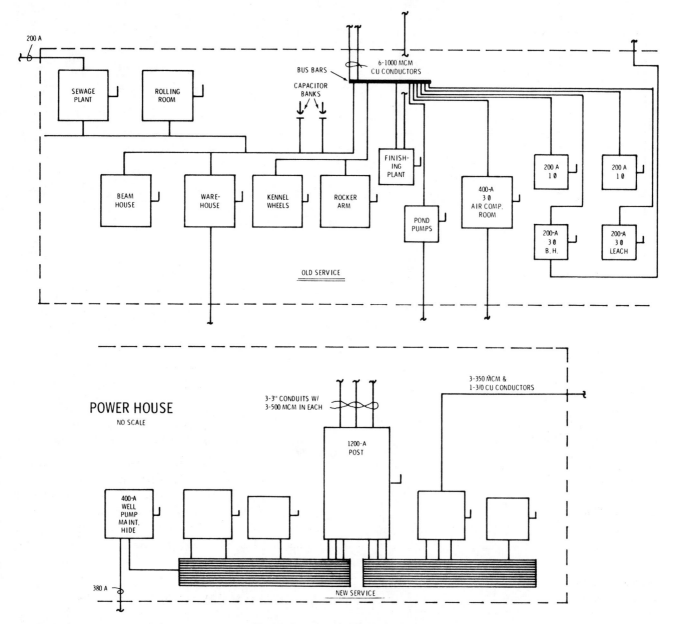

Fig. 1-6. Typical single-line block diagram.

struction of the housing. The second method is to give dimensions on the drawings; this method can be seen in Fig. 1-3. Note that the gauge and type of material are also given in this drawing. The drawing is also drawn to scale.

The drawings of the panelboard housing that was just covered are typical of an electrical construction drawing. They indicate how the equipment will look when completed and show more clearly than any other electrical drawing the actual outlines of equipment installed in their respective locations. All details of the equipment and materials to be used are given on these drawings,

and all dimensions, notes, and references are shown. The complete construction drawing gives all the physical information necesary for installing or erecting the equipment.

Electrical construction drawings, such as the drawing of the panelboard housing in Fig. 1-3, are used by electrical-equipment manufacturers. Electric utility companies also use drawings, such as the one in Fig. 1-4 giving details on the construction of a high-voltage transmission line. A consulting engineering firm may use an electrical construction drawing to supplement building electrical-system drawings for a special installation (Fig.

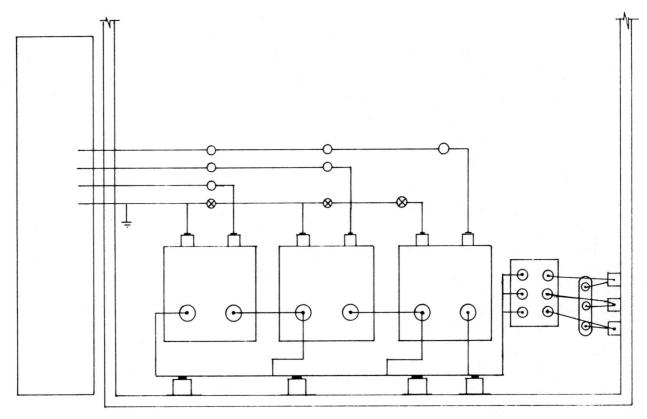

Fig. 1-7. A typical schematic wiring diagram.

Electrical Diagrams

1-5). The latter is often referred to as an *electrical detail drawing*.

Electrical diagrams are drawings that are intended to show, in diagrammatic form, electrical components and their related connections. They are seldom, if ever, drawn to scale, and show only the electrical association of the different components. In diagram drawings, symbols are used extensively to represent the various pieces of electrical equipment, and lines are used to connect these symbols, indicating the size, type, and number of conductors (wires) that are necessary to complete the electrical circuit.

Single-line block diagrams are used extensively by consulting engineering firms to indicate the electric service equipment. The power-riser diagram in Fig. 1-6, for example, is typical of such drawings. The drawing shows all pieces of electrical equipment as well as the connecting lines used to indicate the circuits. Notes are used to identify the equipment, indicate the size of conduit necessary for each circuit, and the number, size, and type of conductors in each conduit. A panelboard schedule usually is included with single-line power-riser dia-

grams to indicate the exact components (fuses, circuit breakers, etc.) contained in each panelboard.

A schematic wiring diagram (Fig. 1-7) is similar to a single-line block diagram except that the schematic diagram gives more-detailed information and the actual number of wires used for the electrical connections are shown.

ELECTRICAL WORKING DRAWINGS

Electrical working drawings that are prepared by architects and consulting engineers to describe the electrical system in a building are very unique drawings. Most drawings encompass all of the previously described types of electrical drawings on each separate building project. For example, a complete set of working drawings for an electrical system usually will consist of the following:

1. A plot plan showing the location of the building on the property and all outside electrical wiring, including the service entrance. This plan is drawn to scale with the exception of various electrical symbols which must be enlarged to be readable.

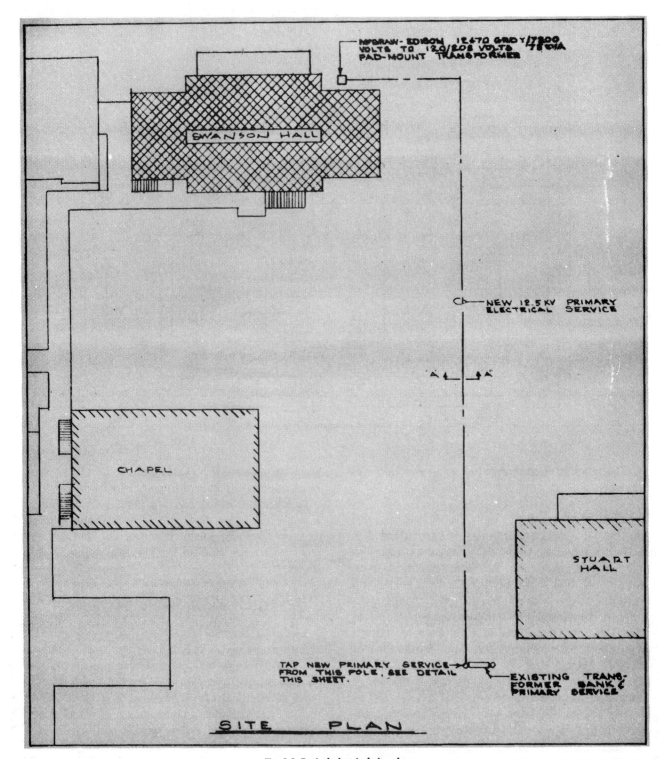

Fig. 1-8. Typical electrical site plan.

Fig. 1-8 shows a typical building plot plan with related electrical wiring.

2. Floor plans showing the walls and partitions for each floor or level. The physical location of all wiring and outlets are shown for lighting, power, signal and communication, special electrical systems, and related electrical equipment. Again, the building partitions are drawn to scale as are

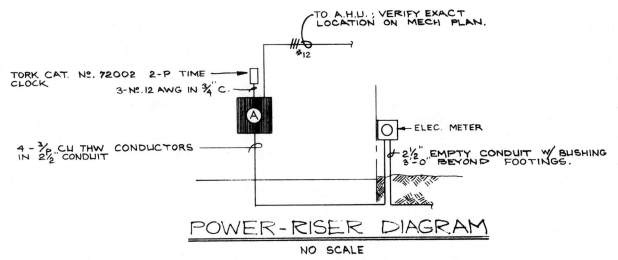

POWER-RISER DIAGRAM

NO SCALE

Fig. 1-10. A typical power-riser diagram.

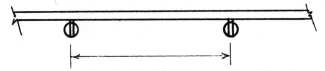

Fig. 1-9. Example of how the spacing of electrical outlets are determined.

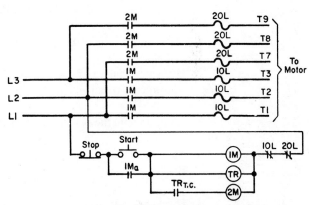

Fig. 1-11. A typical schematic control wiring diagram.

such electrical items as fluorescent lighting fixtures, panelboards, and switchgear. The locations of other electrical outlets and similar components are only approximated on the drawings because they have to be exaggerated. To illustrate, a common duplex receptacle is only about three inches wide. If such a receptacle were to be located on the floor plan of a building that was drawn to a scale of $\frac{1}{8}'' = 1'\text{-}0''$, a small dot on the drawings would be too large to draw the receptacle exactly to scale. Therefore, the symbol ⊖ is used to indicate a duplex receptacle and is clearly shown. When the electrician on the job is locating these outlets, he usually measures to the center of the circle to determine the distance between the outlets (Fig. 1-9).

3. Power-riser diagrams to show the service-entrance and panelboard components (Fig. 1-10).

4. Control wiring schematic diagrams (Fig. 1-11).

PANELBOARD SCHEDULE

PANEL No	TYPE CABINET	PANEL MAINS			BRANCHES					ITEMS FED OR REMARKS
		AMPS	VOLTS	PHASE	1P	2P	3P	PROT	FRAME	
A	FLUSH	200A	120/240 V	3 ∅ 4 W	—	1	—	20A	70A	TIME CLOCK
					—	—	1	20A	70A	A. H. U.
	SQ. "D" TYPE NQOB				—	1	—	30A	70A	WATER HEATER
	W/ MAIN BREAKER				—	—	1	30A	70A	CONDENSING UNIT
					5	—	—	20A	70A	LIGHTS
					10	—	—	20A	70A	RECEPTS
					5	—	—	20A	70A	SPARES
					12	—	—	—	—	PROVISIONS ONLY

Fig. 1-12. A typical panelboard schedule.

5. Schedules, notes, and large-scale details on construction drawings (Fig. 1-12).

In order to be able to "read" an electrical drawing, one must become familiar with the meaning of symbols, lines, and abbreviations used on the drawings and learn how to interpret the message conveyed by the drawings.

SUMMARY

A blueprint or an electrical drawnig is an abbreviated language for conveying a large amount of exact, detailed information, which would otherwise take many pages of manuscript or hours of verbal instruction to convey.

Types of electrical drawings usually fall into the following categories:

1. Electrical construction drawings.
2. Single-line block diagrams.
3. Schematic wiring diagrams.

Electrical working drawings for building construction normally utilize all of the previously described types of electrical drawings.

An electrical working drawing for building construction usually will consist of: a plot plan, floor plans, sectional drawings, various details, wiring diagrams, and schedules.

ASSIGNMENT 1

1. **Describe in your own words an electrical blueprint.**

2. **Drawings that represent the physical arrangement and views of specific electrical equipment are called _____ . _____ drawings.**

3. **Name two methods of indicating the size of an object that is shown on a drawing.**
   A. _____
   B. _____

4. **Drawings that are intended to show, in diagrammatic form, electrical components and their related connections are called _____ .**

5. **A complete set of electrical working drawings usually will consist of:**
   A. _____ plan
   B. _____ plans
   C. _____ diagrams
   D. _____ diagrams

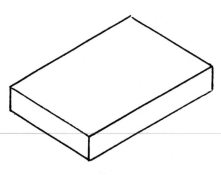

SCALE 1/4" = 1' - 0"

Fig. 1-13. Sketch for assignment.

E. _____
F. _____
G. _____ details

6. The three categories in which most electrical drawings fall are:
A. _____
B. _____
C. _____

7. The sketch in Fig. 1-13 shows a pictorial view of a concrete transformer pad. Redraw (freehand) the sketch in a plan (top, front, and side view).

2

Layout of Electical Drawings

The ideal electrical drawing should show in a clear, concise manner exactly what is required of the workmen. The amount of data shown on such a drawing should be sufficient, but not overdone. This means that a complete set of electrical drawings could consist of only one 8½-inch by 11-inch sheet or it could consist of several dozen sheets, depending on the complexity of the given project.

PREPARING DRAWINGS

Before any drawing is started, the procedure for making the drawing is studied in order to decide what views will be necessary and how they will be arranged. The views that are selected give the proper information for the particular project at hand. All views that are necessary should be shown, but no more.

For example, an electrical design firm has been commissioned to provide working drawings for a water-pumping station. The electrical designer was provided

with a floor plan and elevation views of the pumping-station vault, the location of two 10-horsepower (hp) pumps, and design criteria stating that an electrical service was required along with lighting and power outlets.

Although the designer decided that a drawing of 11 inches by 17 inches would be sufficient to show all necessary details required for the construction of the electrical system, the owners specified that the drawing be made on 24-inch by 36-inch tracing paper in order to fit their standard working-drawing sets.

With this in mind, an electrical draftsman began the working drawings by placing heavy border lines one-half inch inside the outside perimeter of the drawing sheet, as shown in Fig. 2-1. These are heavy solid lines. An outline for a title block was also drawn, to be filled in at a later date.

The designer then decided upon a floor-plan view of the vault showing the location of the wall partitions and door. Within this vault area, the location of the two pumps are shown by means of broken lines (Fig. 2-2). This is to indicate the location of the pumps and to show

Fig. 2-1. Sheet of tracing paper with border lines drawn.

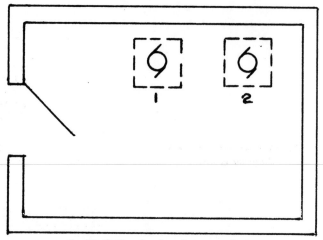

Fig. 2-2. Outline drawing of pump vault plan.

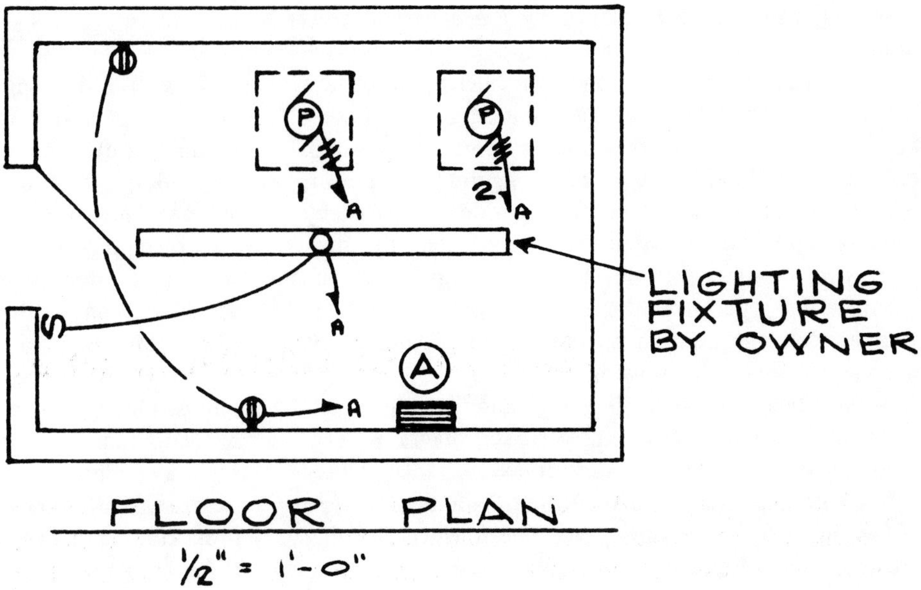
P
1
2
A
S
LIGHTING FIXTURE BY OWNER
A
FLOOR PLAN
1/2" = 1'-0"

Fig. 2-3. Completed floor plan of electrical layout for pumping station.

PANELBOARD SCHEDULE

PNL. No	TYPE CABINET	MAINS AMPS	MAINS VOLT	MAINS PHASE	BRANCHES 1-P	BRANCHES 3-P	BRANCHES PROT	REMARKS
A	SURFACE	100	120/208	3P4W	1	-	20	LTS.
SQ. "D" TYPE NQO W/ 100-A. MAIN					1	-	20	RECEPTS
						2	30	PUMPS

Fig. 2-4. Panelboard schedule for the pumping station.

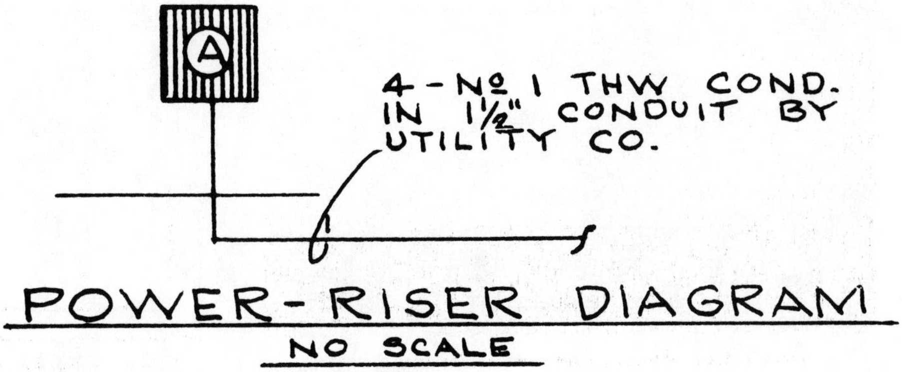
A
4-No 1 THW COND. IN 1½" CONDUIT BY UTILITY CO.
POWER-RISER DIAGRAM
NO SCALE

Fig. 2-5. Power-riser diagram for the pumping station.

that they will not be furnished by the electrical contractor.

Next, the lighting-fixture and convenience outlets are located and circuited, and the location of the panelboard is selected. Since this project is extremely small, only one panelboard is required for lighting and power. The details of the completed floor plan as prepared by the draftsman are shown in Fig. 2-3. Notice that this plan also shows the feeder to each pump. Since the pumps are shipped from the manufacturer as a package with all controls built in, no additional wiring details need be shown on the drawings.

A panelboard schedule (Fig. 2-4) describes the panelboard components, and a power-riser diagram (Fig. 2-5) shows the details of the service entrance. Notes lettered on the drawing give further data on the type of lighting fixtures required, etc. The finished drawing now appears in Fig. 2-6.

ARCHITECT'S SCALE

When the drawings are being laid out, the scale decided upon is very important. Where dimensions must be held to extreme accuracy, the scale drawings should be as large as practical with dimension lines added. Where dimensions require only reasonable accuracy, the object may be drawn to a smaller scale (with dimension lines possibly omitted) since the object can be scaled with an architect's scale.

In most electrical drawings for building construction, the electrical components are so large that it would be impossible to draw them full size. Thus, the drawing is made to some reduced scale—that is, all the distances are drawn smaller than the actual dimensions of the object itself, all dimensions being reduced in the same proportion. For example, if a floor plan of a building is to be drawn to a scale of ¼″ = 1′-0″, each ¼ inch on

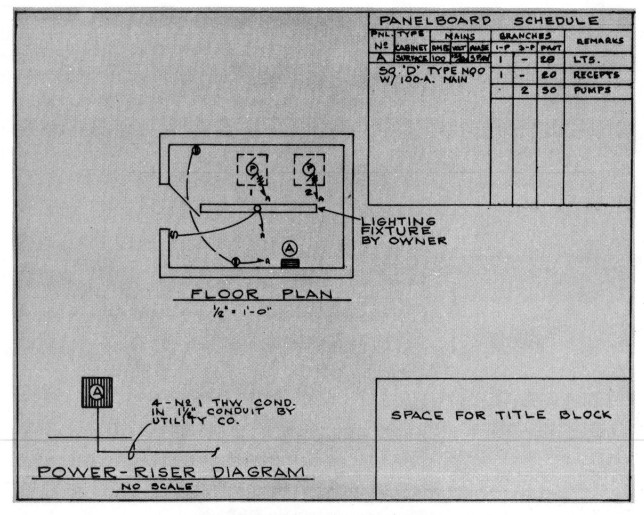

Fig. 2-6. Finishing working drawing for the pumping station.

1″ = 1′-0″.

Fig. 2-7. Typical graduations on an architect's scale.

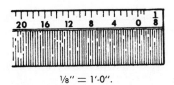

⅛″ = 1′-0″.

Fig. 2-8. Example of larger graduations on the architect's scale.

the drawing would equal 1 foot on the building itself; if the scale was ⅛″ = 1′-0″, each ⅛ inch on the drawing would equal 1 foot on the building, and so forth.

In dimensioning such a drawing, the dimension written on the drawing is the actual dimension of the building, not the distance that is measured on the drawing. To further illustrate this point, look at the floor plan drawing in Fig. 2-6; it is drawn to a scale of ½″ = 1′-0″. One of the walls is drawn to an actual length of 3½ inches on the drawing. Since the scale is ½″ = 1′-0″ and since 3½ inches contains 7 halves of an inch (7 × 0.5 =

3½ inches) the dimension shown on the drawing will therefore be 7′-0″.

From the previous example, we may say that the most common method of reducing all the dimensions in the same proportion is to choose a certain distance and let that distance represent one foot. This distance can then be divided into twelve parts, each of which represents an inch. If half inches are required, these twelfths are further subdivided into halves, etc. We now have a scale that represents the common foot rule with its subdivisions into inches and fractions, except that

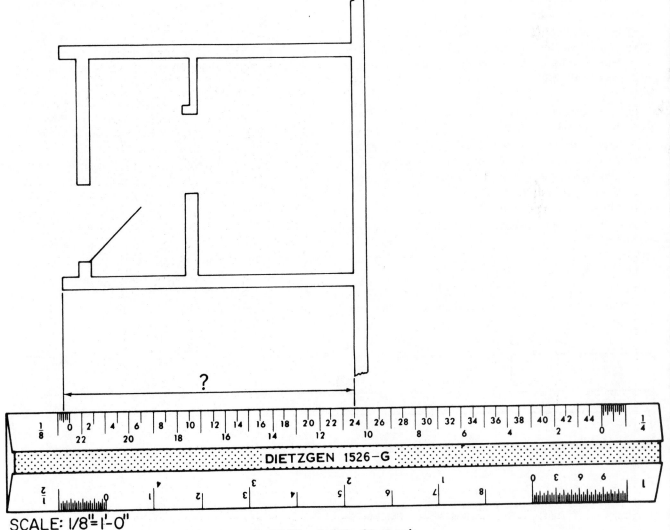

SCALE: 1/8″=1′-0″

Fig. 2-9. Using the ⅛″ architect's scale.

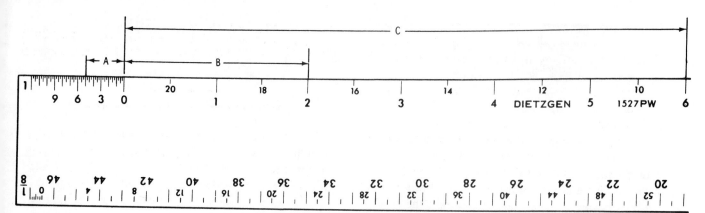

Fig. 2-10. Using the 1" architect's scale.

the scaled foot is smaller than the distance known as a foot and, likewise its subdivisions are proportionately smaller. Therefore, when a measurement is made on the drawing, it is made with the *reduced foot rule;* when a measurement is made on the building, it is made with the *standard foot rule.* The most common reduced foot rules or scales used in electrical drawings are the architect's scale and the engineer's scale. The architect's scale will be fully explained in this chapter, while the engineer's scale will be covered in Chapter 8, "Site Plans."

Fig. 2-7 shows one type of architect's scale. Note that the basic unit at the end of the scale (⅛ in this case) represents one foot and is subdivided into twelve parts to represent inches. On larger scales such as 1" = 1'-0", the inch division is further subdivided so that the smallest subdivision may represent ½ of an inch (Fig. 2-8). On smaller scales, however, the basic unit is not divided into as many divisions. For example, the smallest subdivision on the ⅛" = 1'-0" scale represents 2 inches. The following drawings demonstrate how the various scales are used to determine different lengths.

Fig. 2-9 shows a part of a building floor plan drawn to a scale of ⅛" equals 1'. The dimension in question

is found by placing the ⅛" architect's scale on the drawing and reading the figures. It can be seen that the dimension reads 24'-6".

Fig. 2-10 shows a portion of an architect's scale with the 1" = 1'-0" side turned up. The dimensions of the various lines are as follows: A equals 5", B equals 2'-0", and C equals 6'-0".

If it is desirable to draw a given line to a given scale, first mark off the distance with the appropriate scale; this is indicated by two light dots (Fig. 2-11). Then use a straightedge to draw the line between the dots.

Every drawing should have the scale to which it is drawn, plainly marked on it as part of the Drawing Title, as illustrated in Fig. 2-12. However, it is not uncommon to have several different drawings on one sheet of tracing paper—all with different scales.

In nearly all instances, when a building of any size is planned, an architect is commissioned to plan and design the building. An engineer or electrical designer usually lays out the complete electrical system for the architect's buildings, and an electrical draftsman transforms the engineer's designs into working drawings. In the preparation of the electrical design and working drawings, the following usually takes place:

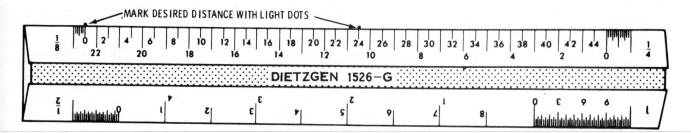

Fig. 2-11. Drawing a given line to a given scale.

1. The engineer meets with the architect and owner to discuss the electrical needs of the building and to discuss various recommendations made by all parties.
2. After that, an outline of the architect's floor plan is laid out.
3. The engineer then calculates the required power and lighting outlets for the building; these are later transferred to the working drawing.
4. All communication and alarm systems are located on the floor plans along with lighting and power panelboards.
5. Circuit calculations are made to determine wire size and overcurrent protection.
6. The main electric service and related components are determined and shown on the drawings.
7. Schedules are then placed on the drawings to identify various pieces of electrical equipment.

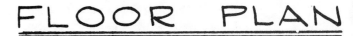

FLOOR PLAN

$\frac{1}{4}'' = 1'-0''$

Fig. 2-12. Example of way to show drawing scale.

8. Wiring diagrams are made to show the workmen how various electrical components are to be connected.
9. A legend or electrical symbol list is shown on the drawings to identify all symbols used to indicate electrical outlets or equipment.
10. Various large-scale electrical details are included, if necessary, to show exactly what is required of the workmen.
11. Written specifications are then made to give a description of the materials and installation methods.

ASSIGNMENT 2

1. Give the missing dimensions shown in the illustration of the architect's scale in Fig. 2-13.

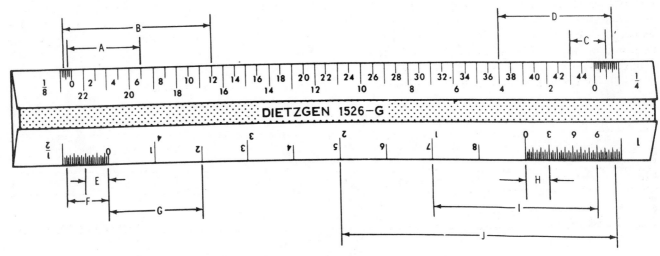

Fig. 2-13. Illustration for assignment problem.

A. _____ B. _____ C. _____ D. _____ E. _____

F. _____ G. _____ H. _____ I. _____ J. _____

2. Using an architect's scale of ⅛″ = 1′-0″, find the missing dimension shown in the illustration of a partial floor plan in Fig. 2-14.

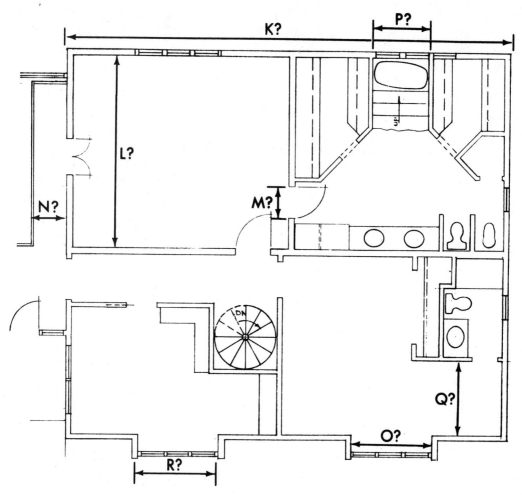

Fig. 2-14. Illustration for assignment problem.

K. _____ L. _____ M. _____ N. _____
O. _____ P. _____ Q. _____ R. _____

3

Electrical Symbols

The purpose of an electrical drawing is to show the location of all lighting and power outlets, the wire sizes, the panelboards, the service equipment, the communications equipment, and other information necesary for the proper construction of an electrical system. In such drawings, symbols are used to simplify the work of those preparing the drawings. In turn, a knowledge of electrical symbols must also be acquired by anyone who must interpret the drawings.

In preparing electrical drawings, most engineers and designers use symbols adopted by the United States of America Standards Institute (USASI). However, many designers frequently modify these standard symbols to suit their own needs. For this reason, most drawings will have a symbol list or legend.

Fig. 3-1 shows a list of electrical symbols currently benig used by one consulting-engineering firm. This list represents a good set of electrical symbols in that they are:

1. Easy for draftsmen to draw.
2. Easily interpreted by workmen.
3. Sufficient for most applications.

It is evident from this list that many of the symbols have the same basic form but their meanings are different because of some slight difference in the symbol.

For example, all the outlet symbols in Fig. 3-2 have the same basic form—a circle; however, the addition of a line or a dot to the circle gives each an individual meaning. It is also apparent that the difference in meaning may be indicated by the addition of letters or an abbreviation to the symbol. Therefore, a good procedure to follow in learning the different symbols is to first learn the basic form and then apply the variations for obtaining different meanings.

Some of the symbols used on electrical drawings are abbreviations, such as "WP" for weatherproof and "AFF" for above finished floor. Others are simplified pictographs, such as ⊘ for a double floodlight fixture or ▭ for an infrared electric heater with two quartz lamps.

In some cases, the symbols are combinations of abbreviation and pictographs, such as [F] for fusible safety switch, [DT]₃₀ for a double-throw safety switch, and [N]₆₀ for a nonfusible safety switch. In each example, a pictograph of a switch enclosure has been combined with an abbreviation—F (fusible), DT (double-throw), and N (nonfusible), respectively. The numerals indicate the bus-bar capacity in amperes.

LIGHTING OUTLETS

The lighting-outlet symbols represent both incandescent and fluorescent types; a circle usually represents an incandescent fixture and a rectangle, a fluorescent one. All of these symbols are designed to indicate the physical shape of a particular fixture and should be drawn as close to scale as possible.

The type of mounting used for all lighting fixtures is usually indicated in a lighting-fixture schedule, which is shown either on the drawings or in the written specification.

The mounting heights of wall-mounted fixtures are usually indicated in the symbol lists, especially where most are to be mounted at one height. For example, a

NOTE: THESE ARE STANDARD SYMBOLS AND MAY NOT ALL APPEAR ON THE PROJECT DRAWINGS; HOWEVER, WHEREVER THE SYMBOL ON PROJECT DRAWINGS OCCURS, THE ITEM SHALL BE PROVIDED AND INSTALLED.

FLUORESCENT STRIP

FLUORESCENT FIXTURE

INCANDESCENT FIXTURE, RECESSED

INCANDESCENT FIXTURE, SURFACE OR PENDANT

INCANDESCENT FIXTURE, WALL-MOUNTED

LETTER "E" INSIDE FIXTURES INDICATES CONNECTION TO EMERGENCY LIGHTING CIRCUIT

NOTE: ON FIXTURE SYMBOL, LETTER OUTSIDE DENOTES SWITCH CONTROL.

EXIT LIGHT, SURFACE OR PENDANT

EXIT LIGHT, WALL-MOUNTED

INDICATES FIXTURE TYPE

RECEPTACLE, DUPLEX-GROUNDED

RECEPTACLE, WEATHERPROOF

COMBINATION SWITCH AND RECEPTACLE

RECEPTACLE, FLOOR-TYPE

RECEPTACLE, POLARIZED (POLES AND AMPS INDICATED)

SWITCH, SINGLE-POLE

SWITCH, THREE-WAY, FOUR-WAY

SWITCH AND PILOT LIGHT

SWITCH, TOGGLE W THERMAL OVERLOAD PROTECTION

PUSH BUTTON

BUZZER

LIGHT OR POWER PANEL

TELEPHONE CABINET

JUNCTION BOX

DISCONNECT SWITCH – FSS – FUSED SAFETY SWITCH. NFSS – NONFUSED SAFETY SWITCH

STARTER

ABOVE FINISHED FLOOR

CONDUIT, CONCEALED IN CEILING OR WALL

CONDUIT, CONCEALED IN FLOOR OR WALL

CONDUIT, EXPOSED

FLEXIBLE METALLIC ARMORED CABLE

HOME RUN TO PANEL – NUMBER OF ARROWHEADS INDICATES NUMBER OF CIRCUITS. NOTE: ANY CIRCUIT WITHOUT FURTHER DESIGNATION INDICATES A TWO-WIRE CIRCUIT. FOR A GREATER NUMBER OF WIRES, READ AS FOLLOWS – ─///─ 3 WIRES, ─////─ 4 WIRES, ETC.

TELEPHONE CONDUIT

TELEVISION – ANTENNA CONDUIT

SOUND-SYSTEM CONDUIT – NUMBER OF CROSSMARKS INDICATES NUMBER OF PAIRS OF CONDUCTORS.

FAN COIL-UNIT CONNECTION

MOTOR CONNECTION

MOUNTING HEIGHT

FIRE-ALARM STRIKING STATION

FIRE-ALARM GONG

FIRE DETECTOR

SMOKE DETECTOR

PROGRAM BELL

YARD GONG

CLOCK

MICROPHONE, WALL-MOUNTED

MICROPHONE, FLOOR-MOUNTED

SPEAKER, WALL-MOUNTED

SPEAKER, RECESSED

VOLUME CONTROL

TELEPHONE OUTLET, WALL

TELEPHONE OUTLET, FLOOR

TELEVISION OUTLET

Fig. 3-1. Electrical symbols used by a consulting engineering firm.

DUPLEX
RECEPTACLE

COMBINATION SWITCH
& RECEPTACLE

DUPLEX RECEPTACLE WITH
WATERPROOF COVER

RECEPTACLE,
FLOOR TYPE

Fig. 3-2. Example of various receptacle symbols.

Fig. 3-3. Example of showing a wall-mounted lighting fixture on an electrical drawing.

motel project may have 100 rooms with a wall-mounted lighting fixture outside of each room door. The symbol lists might read, ". . . wall outlet with incandescent fixture mounted 6 feet, 6 inches above finished floor to center of outlet box unless otherwise indicated." If a few other wall-mounted fixtures on this project needed to be mounted at a different height, they could be indicated as shown in Fig. 3-3.

The type of lighting fixture is identified by a numeral placed inside a triangle near each lighting fixture, as shown in Fig. 3-3. If one type of fixture is used exclusively in one room or area, such as in Fig. 3-4, the indicator need only appear once with the word "all" lettered at the bottom of the indicator triangle.

A complete description of the fixture identified by the symbol must be given in the lighting-fixture schedule and should include the manufacturer, catalog number, number and type of lamps, voltage, finish, mounting, and any other information needed for a proper installation of the fixtures. Chapter 7 gives examples of lighting-fixture schedules.

SWITCHES AND RECEPTACLES

A single-pole switch is used to control lighting from one point, while three- and four-way switches are used

in combination to control a single light or a group of lights from two or more points.

A two-pole switch is used to control a series of lights on two separate circuits with only one motion or to control single-phase 240-volt loads. The switch and pilot-light combination is used where it is practical—if the lighting fixture controlled by the switch is located in a closet, basement, or attic space.

Door switches in residential construction are commonly used to control closet lights. The operation of these switches is very simple: when the door is closed, the bottom of the switch (located in the door jamb) is depressed, which opens the circuit; when the door is opened, the switch button is released—closing the circuit—and the light comes on.

Table 3-1. Electrical Abbreviations

CSP	Central switch panel
MDP	Main distribution panel
DCP	Dimmer control panel
DT	Dust tight or double throw
ESP	Emergency switch panel
MT	Empty
EP	Explosion proof
Gd	Grounded
NL	Night light
PC	Pull chain
RT	Raintight
R	Recessed
SFER	Transfer
SFRMR	Transformer

Fig. 3-4. Lighting plan illustrating the need of only one fixture-type indicator if all fixtures are of the same type.

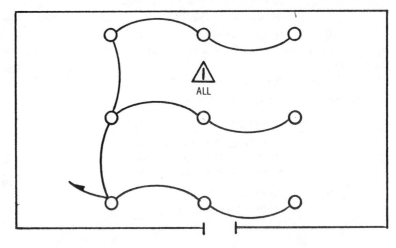

While the types of receptacle symbols used in consulting engineering firms are numerous, the few shown in the symbol list should suffice for most electrical drawings used for building construction. If other symbols are needed to indicate various outlets on drawings, they may be composed and added to the symbol list. A description of their use must always be included.

When outlets are located in areas requiring special outlet boxes, covers, fittings, etc., they are usually indicated by abbreviations such as WP (weatherproof), EP (explosion proof), etc. (See Table 3-1 for a list of abbreviations.)

SERVICE EQUIPMENT, FEEDERS, AND BRANCH CIRCUITS

Main distribution centers, panelboards, transformers, safety switches, and other similar electrical components are indicated by electrical symbols on floor plans and by a combination of symbols and semipictorial drawings in riser diagrams.

A detailed description of the service equipment is usually given in the panelboard schedule (Chapter 7) or in the written specifications. However, on small projects, the service equipment is sometimes indicated only by notes on the drawings.

Circuit and feeder wiring symbols have been nearly standarized. Most circuits concealed in the ceiling or wall are indicated by a solid line; a broken line is used for circuits concealed in the floor or ceiling below; and exposed raceways are indicated by short dashes.

The number of conductors in a conduit or raceway system may be indicated in the panelboard schedule under the appropriate column, or the information may be shown on the floor plan.

For example, the symbol list states that a circuit line with no slash marks or numerals indicates a circuit containing two No. 12 AWG conductors. Three slash marks with no numeral indicates three No. 12 AWG conductors, etc. If the circuit contained two No. 10 AWG conductors, then two slash marks would be used along with the numeral 10.

Most electrical drawings use an arrowhead on the end of a circuit to indicate a "home run" to the panelboard. The number of arrowheads indicate the number of circuits in the run. Again, the slash marks indicate the number of conductors in the run.

COMMUNICATION AND ALARM SYMBOLS

Symbols for communication and signal-systems as well as symbols for light and power, are drawn to an appropriate scale and accurately located with respect to the building; this reduces the number of references made to the architectural drawings. Where extremely accurate final location of outlets and equipment is necessary, exact dimensions are given on larger-scale drawings and shown on the plans.

Each different category in a signal system is usually represented by a distinguishing basic symbol. Every item of equipment or outlet in that category of the system is identified by that basic symbol. To further identify items of equipment or outlets in the category, a numeral or other identifying mark is placed within the open basic symbol. In addition, all such individual symbols used on the drawings should be included in the symbol list or legend.

SUMMARY

In electrical drawings, symbols are used to simplify the work of those preparing the drawings and to make the interpretation of the drawings less complex.

Most engineers and designers use standard electrical symbols adapted by the United States of America Standards Institute (USASI).

A good procedure to follow in learning different symbols is to first learn the basic form and then apply the variations for obtaining different meanings.

All symbols used on scale drawings are drawn to scale, when practical, and are accurately located with respect to the building. This is in order to reduce the number of references made to the architectural drawings.

ASSIGNMENT 3

Shown below are 20 symbols commonly found on electrical working drawings. In the space provided, place the letter corresponding to the correct answer found in the list.

1. ⊝━   _____   A. **Fluorescent fixture**

2. ⊗   _____   B. **Incandescent fixture, recessed**

3. ⊂◯wp   _____   C. **Incandescent fixture, wall-mounted**

4. ☐F   _____   D. **Exit light, surface- or pendant-mounted**

5. ─ ─ ─ ─   _____   E. **Exit light, wall-mounted**

6. ─⫴→   _____   F. **Indicates fixture type**

7. s⊝━   _____   G. **Receptacle, duplex-grounded**

8. ☐◯☐   _____   H. **Receptacle, weatherproof**

9. ◯┤   _____   I. **Combination switch and receptacle**

10. (SD) _____   J. **Receptacle, floor-type**

11. ▶| _____   K. **Switch, three-way**

12. ☐┘   _____   L. **Light or power panel**

13. ▬━   _____   M. **Disconnect switch**

14. ⊙   _____   N. **Conduit, exposed**

15. S₃   _____   O. **Home run to panel**

16. △A   _____   P. **Telephone conduit**

17. ⊢⊗   _____   Q. **Fan coil-unit connection**

18. ☐   _____   R. **Fire-alarm striking station**

19. ─T─   _____   S. **Smoke detector**

20. ◯F   _____   T. **Telephone outlet, wall**

Types of Building Drawings

The four basic types of building drawings found in a set of electrical working prints are:

1. Plans.
2. Elevations.
3. Sections.
4. Details.

Plans are viewed directly above; elevations are head-on views of vertical surfaces, and sections are "cut" or "sliced-open" views that show the actual composition of the part of the building under consideration. Together with the written specifications, these documents must give all details of construction so that the workmen will know exactly what is required of them for proper construction of the building or system. Floor plans are the most commonly used type of drawing for showing the location of electrical outlets, equipment, and components of an electrical system within a building.

A drawing, known as an orthographic projection of a building, is the most frequently used type of drawing for all the views listed previously. For example, a pictorial drawing (perspective of a building) is shown in Fig. 4-1. While this drawing shows how the building will appear to viewers after it is completed, it would be difficult, indeed, to construct the building from this type of drawing because all the sides of the building are not shown, nor is the interior of the building. An ortho-

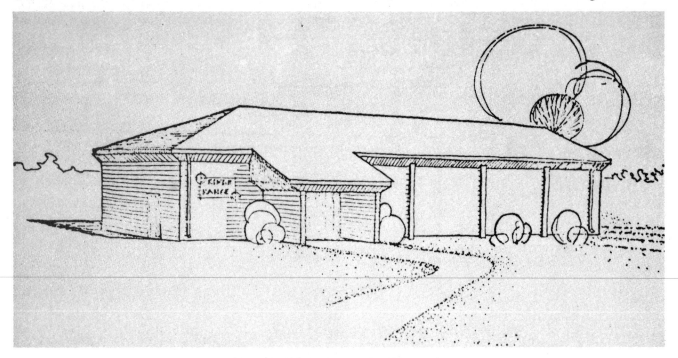

Fig. 4-1. Perspective drawing of a small commercial building.

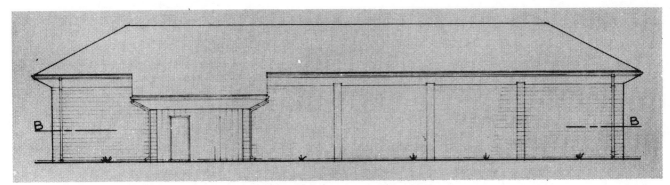

Fig. 4-2. Front elevation of the building in Fig. 4-1.

graphic-projection drawing of this same building would consist of the views illustrated in Figs. 4-2, 4-3, and 4-4. Fig. 4-2 shows the building as though the observer were looking straight at the front. Fig. 4-3 shows a straight view of the rear of the building. Fig. 4-4 shows the side views of the building. These drawings are called building elevations.

Referring again to Fig. 4-2, assume that a horizontal cut is made through the building along a line indicated as B-B on the front elevation. Then, imagine that the top part of the building is removed and a drawing is made looking straight down at the remaining part. This drawing is known as a floor plan. It indicates the outside wall lines, interior partitions, windows, doors, and similar details. Fig. 4-5 shows a drawing of the floor plan for the building illustrated in Fig. 4-2.

The building in Figs. 4-1 and 4-2 has only one floor. Should a building have more than one floor, horizontal cuts would be taken at varying distances from the ground in order to show the other floors. Also, if the original scale of the overall floor plan is too small to clearly show all necessary details of construction, a portion of the building may be drawn to a larger scale (Fig. 4-6) in order to solve this problem.

Fig. 4-7 shows a sample floor plan from a set of electrical working drawings. Referring to the electrical

symbols in Chapter 3, where necessary, the floor plan is analyzed as follows.

First, notice that the entire drawing sheet has a border line. This is to square and confine the drawings. A title block identifies the project as well as the architect who prepared the drawings. Each drawing on this sheet is also titled; that is, each has a subtitle as opposed to the job title shown in the title block. The "Lighting-Fixture Plan" is indicated as being drawn to a scale of ¼″ = 1′-0″.

The floor plan of this building shows all outside walls, interior partitions, windows, doors, and toilet fixtures. It can be seen that most of the ceiling (from the reflected ceiling grid) consists of an inverted T-bar ceiling with 2-foot by 4-foot lay-in tiles or panels. All of these items are installed by contractors other than the electrical contractor and are therefore drawn on the plan slightly lighter than the electrical components, which will now be described.

LIGHTING FIXTURES

From the symbol list in Chapter 3, we can see that twenty-nine 2-foot by 4-foot fluorescent lighting fixtures are the predominant source of light for the entire building. All of these fixtures are identified as type "1." With

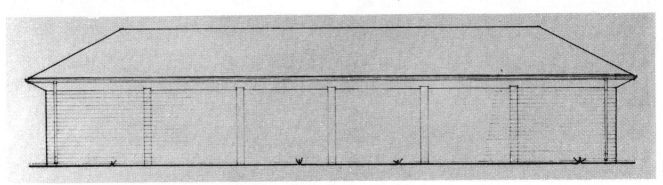

Fig. 4-3. Rear elevation of the building in Fig. 4-1.

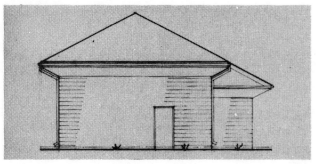

(A) Left-side elevation.

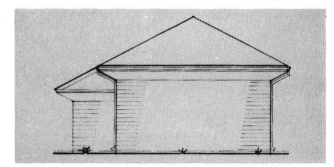

(B) Right-side elevation.

Fig. 4-4. Side elevations of building in Fig. 4-1.

this information, the type "1" fixture is found in the "Lighting-Fixture Schedule," which in turn gives the reader additional information. The fixture is manufactured by Benjamin, and the catalog number is AG-7244-4. Continuing across the row horizontally, we know that the fixture contains four 40-watt lamps, is rated at 120 volts, and is to be recessed in the ceiling. Since the ceiling is of the T-bar type, an experienced electrician will know that the type "1" fixture will be laid in the ceiling grid in a manner similar to the ceiling panels. This type of lighting fixture is known as a *lay-in troffer*.

The electrical symbol for the type "2" lighting fixture identifies it as a fluorescent fixture also. Again, refering to the lighting-fixture schedule, we see that it is manufactured by Crescent; its catalog number is ANG 220; it contains two, 20-watt fluorescent lamps rated at 120 volts; and it is wall mounted above the lavatory mirror in the toilets.

Five other lighting-fixture types are shown on the floor plans and then identified by their corresponding numbers in the lighting-fixture schedule. The letter *F*

in the "Lamps" column indicates that the lamps are fluorescent, and the letter *I* indicates that the lamps are incandescent. All of the lighting fixtures are rated for use on 120-volt circuits. The vertical line drawn down the "Volts" column indicates that the 120-volt figure is valid for all the rows; this method is used rather than repeat the 120-volt figure in every row. Additional information about the electrical schedule may be found in Chapter 7.

LIGHTING CIRCUITS

Since all of the branch circuits feeding the lighting fixtures are concealed in either the ceiling or the wall, all are shown with a solid line (this is the conventional symbol). The half arrowheads on the lines indicate a homerun to the panelboard, which in this case is panel A.

Circuit No. 22 feeds eight type-1 lighting fixtures installed in two rows of four fixtures. Since the four fixtures are butted together, the wiring is fed through the fixtures themselves and is therefore not shown on the

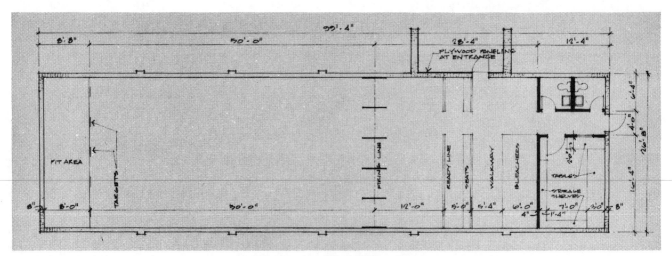

Fig. 4-5. Floor plan of the building in Fig. 4-1.

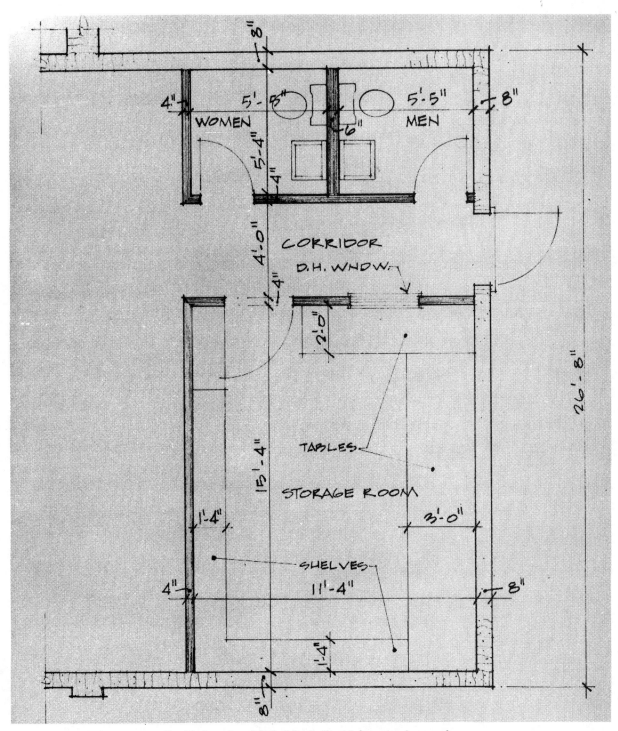

WOMEN

MEN

CORRIDOR

D.H. WNDW.

TABLES

STORAGE ROOM

SHELVES

Fig. 4-6. A portion of the building in Fig. 4-1 drawn to a larger scale.

drawings. However, a circuit is shown connecting the two rows together, and then another circuit is shown with an arrowhead to indicate that the circuit goes directly to panel A. The "A-22" adjacent to the arrowhead indicates that the circuit connects to the No. 22 overcurrent-protection device (circuit breaker) in panel A.

Circuit No. 24 feeds four type-7 lighting fixtures, two type-2 fixtures, four type-5 fixtures, and two type-4 fixtures. Since only two wires (conductors)

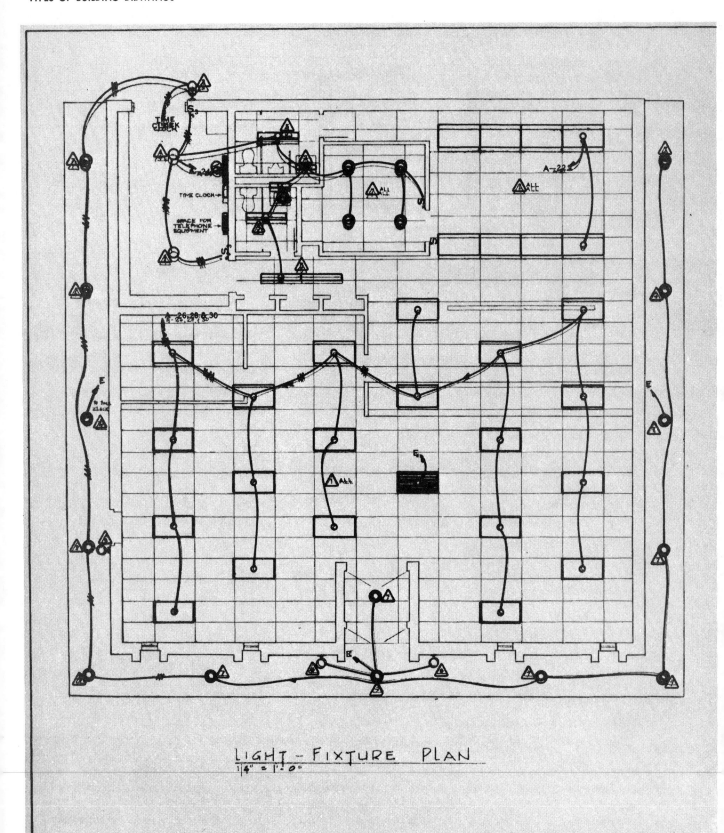

LIGHT — FIXTURE PLAN
1/4" = 1'-0"

Fig. 4-7. Lighting layout of

LIGHTING-FIXTURE SCHEDULE

FIXT TYPE	MANUFACTURER'S DESCRIPTION	LAMPS NO.	LAMPS TYPE	VOLTS	MOUNTING	REMARKS
⚠1	BENJAMIN CAT. No. AG-7244-4	4	40W F	120	RECESSED	
⚠2	CRESCENT CAT. No. ANG 220	2	20W F		WALL	
⚠3	MOLDCAST CAT. No. 2100	1	150W I		WALL	
⚠4	BENJAMIN CAT. No. 9642	2	150W I		SURFACE	
⚠5	BENJAMIN CAT. No. CD-2214-4	1	40W F		SURFACE	
⚠6	MOLDCAST CAT. No. 531	1	150W I		WALL	
⚠7	MOLDCAST CAT. No. A-270	1	150W I		RECESSED	
⚠8						
⚠9						
⚠10						

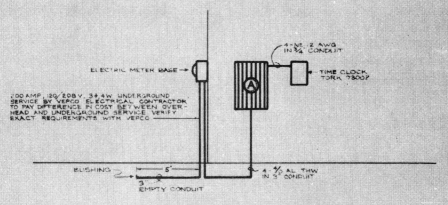

ELECTRIC METER BASE →

4-No. 2 AWG IN ¾" CONDUIT

TIME CLOCK TORK 7300Z

200 AMP, 120/208 V, 3Φ, 4W UNDERGROUND SERVICE BY VEPCO ELECTRICAL CONTRACTOR TO PAY DIFFERENCE IN COST BETWEEN OVERHEAD AND UNDERGROUND SERVICE VERIFY EXACT REQUIREMENTS WITH VEPCO

BUSHING

5'

EMPTY CONDUIT

4-⅘ AL THW IN 3" CONDUIT

POWER-RISER DIAGRAM

NO SCALE

THESE PLANS ARE THE SOLE PROPERTY OF THE ARCHITECT AND MAY NOT BE USED FOR OTHER PROJECTS EXCEPT BY WRITTEN PERMISSION OF THE ARCHITECT.

COMMONWEALTH OF VIRGINIA
G. LEWIS CRAIG
CERTIFICATE No.
1698
CERTIFIED ARCHITECT

LIGHT-FIXTURE PLAN & SCHEDULE

DMV BRANCH OFFICE
FISHERSVILLE, VIRGINIA

G. LEWIS CRAIG, ARCHITECT
WAYNESBORO   VIRGINIA

COMM. NO	DATE	DRAWN	CHECKED	REVISED
7411	6-25-74	JET	GLC	

SHEET NO. E-1

a commercial building.

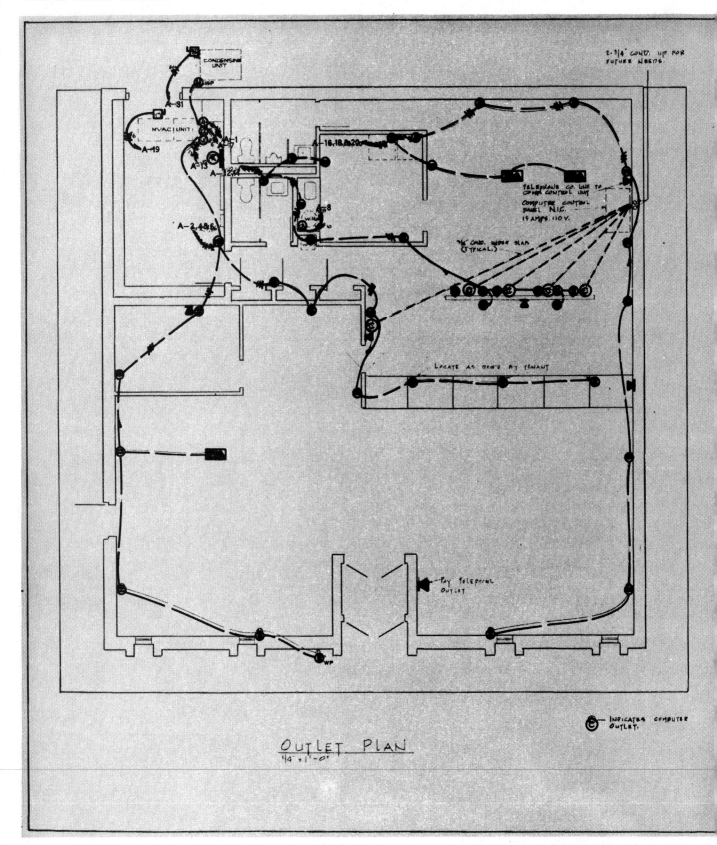

OUTLET PLAN
1/4" = 1'-0"

Ⓒ — INDICATES COMPUTER OUTLET.

Fig. 4-8. Power layout drawing of

ELECTRICAL SYMBOL LIST

NOTE: THESE ARE STANDARD SYMBOLS AND MAY NOT ALL APPEAR ON THE PROJECT DRAWINGS. HOWEVER, WHEREVER THE SYMBOL ON PROJECT DRAWINGS OCCURS, THE ITEM SHALL BE PROVIDED AND INSTALLED.

- CEILING OUTLET WITH INCANDESCENT FIXTURE
- RECESSED OUTLET WITH INCANDESCENT FIXTURE
- WALL-MOUNTED OUTLET WITH INCANDESCENT FIXTURE
- CEILING OUTLET WITH FLUORESCENT FIXTURE
- WALL-MOUNTED OUTLET WITH FLUORESCENT FIXTURE
- FLUORESCENT FIXTURE
- FLUORESCENT FIXTURE MOUNTED UNDER CABINET
- GROUND-MOUNTED UPLIGHT
- POST-MOUNTED INCANDESCENT FIXTURE
- FLOOD LIGHT FIXTURE
- FLOOD LIGHT FIXTURE
- FLUORESCENT STRIP
- EXIT LIGHT, SURFACE OR PENDANT
- EXIT LIGHT, WALL MOUNTED
- INDICATES TYPE OF LIGHTING FIXTURE — SEE SCHEDULE
- SINGLE-POLE SWITCH MOUNTED 50" UP TO ₵ OF BOX
- THREE-WAY SWITCH MOUNTED 50" UP TO ₵ OF BOX
- FOUR-WAY SWITCH MOUNTED 50" UP TO ₵ OF BOX
- TWO-POLE SWITCH MOUNTED 50" UP TO ₵ OF BOX
- LOW VOLTAGE SWITCH TO RELAY
- DOOR SWITCH
- DUPLEX RECEPTACLE MOUNTED 18" UP TO CENTER OF BOX

- DUPLEX RECEPTACLE MOUNTED 4" ABOVE COUNTERTOP
- SPLIT-WIRED DUPLEX RECEPTACLE - TOP HALF SWITCHED
- SPECIAL OUTLET OR CONNECTION - NUMERAL INDICATES TYPE SEE LEGEND AT END OF SYMBOL LIST
- FLOOR-MOUNTED RECEPTACLE
- CLOCK HANGER RECEPTACLE
- PUSHBUTTON SWITCH FOR DOOR CHIMES
- CHIMES
- TV OUTLET MOUNTED 18" UP TO ₵ OF BOX
- TELEPHONE OUTLET
- FUSIBLE SAFETY SWITCH
- NON-FUSIBLE SAFETY SWITCH
- MAIN DISTRIBUTION PANEL
- LIGHTING PANEL NUMERAL INDICATE TYPE
- BRANCH CIRCUIT CONCEALED IN CEILING OR WALLS SLASH MARKS INDICATE NUMBER OF CONDUCTORS IN RUN, TWO CONDUCTORS NOT NOTED
- BRANCH CIRCUIT CONCEALED IN FLOOR OR CEILING BELOW
- LOW VOLTAGE CABLE
- INDICATES TYPE OF HEATER — SEE SCHEDULE
- INDICATES HOMERUN TO PANELBOARD - NUMBER OF ARROW HEADS INDICATES NUMBER OF CIRCUITS

- WEATHERPROOF
- MOTOR OUTLET, NUMERAL INDICATES H.P.
- JUNCTION BOX
- DIMMER CONTROL FOR LIGHTING FIXTURE
- ELECTRIC BASEBOARD HEATER
- FLUSH MOUNTED ELECTRIC FLOOR HEATER
- CEILING ELECTRIC PANEL HEATER
- INFRA-RED ELECTRIC HEATER — CEILING-MOUNTED
- DOUBLE-POLE THERMOSTAT FOR ELECTRIC HEAT
- FIRE ALARM STRIKING STATION
- FIRE-ALARM GONG
- FIRE DETECTOR
- SMOKE DETECTOR
- PROGRAM BELL
- YARD GONG
- MICROPHONE, WALL-MOUNTED
- MICROPHONE, FLOOR-MOUNTED
- SPEAKER, WALL-MOUNTED
- SPEAKER, RECESSED

CCT NO	VOLT-AMPERES			DESCRIPTION	OUTLETS	CCT BKR	PHASE A B C	CCT BKR	OUTLETS	DESCRIPTION	VOLT-AMPERES			CCT NO
	φA	φB	φC								φA	φB	φC	
1	5333			ELEC. RESIST. HTR		3 60				RECEPTS	1500			2
3		5333								RECEPTS		1200		4
5			5333							RECEPTS			1200	6
7	5333			ELEC. RESIST. HTR		3 60				WATER HEATER	2250			8
9		5333										2250		10
11			5333							RECEPTS			1800	12
13	5333			ELEC. RESIST. HTR		3 60				RECEPTS	1200			14
15		5333								RECEPTS		1200		16
17			5333							RECEPTS			1200	18
19	800			AIR HANDL'G UNIT		3 20				RECEPTS	1200			20
21		800								LIGHTS		1600		22
23			800							LIGHTS			1350	24
25	1500			TIME CLOCK		3 20								26
27		1200												28
29			100											30
31	5316			COND. UNIT		3 60								32
33		5316												34
35			5316											36
	23615	23315	22115			SUB-TOTALS					6150	5250	5550	
	29765	28565	27665											

PANEL A 120/208 V 3φ 4 WIRE SURFACE MOUNTED 200 AMPERE MAIN - BREAKER
LOCATION MECHANICAL ROOM 200 AMPERE BUS

TOTAL VA/φ
LCL ADDER
TOTAL VA
LINE AMPS

SQ "D" TYPE QOB W/MAIN BREAKER

COMMONWEALTH OF VIRGINIA
G. LEWIS CRAIG
CERTIFICATE No.
1698
CERTIFIED ARCHITECT

OUTLET PLAN - PANEL HOOKUP
DMV BRANCH OFFICE
FISHERSVILLE, VIRGINIA.

G. LEWIS CRAIG, ARCHITECT
WAYNESBORO   VIRGINIA

COMM. NO.	DATE	DRAWN	CHECKED	REVISED
7411	6-27-74	JET	CLC	

SHEET NO.
E-2

the building in Fig. 4-7.

will be required for most of the circuit, slash marks are not required. However, the two type-4 lighting fixtures are controlled by two 3-way switches which require extra conductors in the raceway for "traveler" wires. From each 3-way switch (S_3) to the lighting fixtures, each circuit contains three conductors, and this is indicated by three slash marks through the circuit lines. Four conductors are required between the two fixtures and, again, are indicated by the appropriate number of slash marks.

Three separate circuits are necessary to feed the remaining type 1 fixtures in the building. However, the three circuits are combined in one raceway or conduit and use only one common neutral wire for all three circuits; three circuits may be used with one common neutral on three-phase systems, and two circuits with one common neutral are permitted on single-phase systems.

The first circuit feeds eight type-1 fixtures and then runs to an outlet box to combine with the second circuit which feeds five type-1 fixtures. At this point, three conductors are necessary to carry both circuits to the third circuit; that is, two "hot" conductors and one common neutral. Notice that an arrowhead indicates where each circuit ends.

A single type-1 fixture is different from the rest in that it is connected to the emergency panel (E) and is used as an emergency white light which burns all the time. Two type-7 fixtures, recessed in the soffit of the building, and one type-7 fixture on each end of the building are also connected to the emergency panel and will burn continuously. The remaining type-7 fixtures on the outside of the building are fed by two circuits that are connected to a time clock or timer switch. This clock is adjusted to turn the exterior lights on at dusk and then turn them off again at dawn or any other time the owners choose. The accompanying power-riser diagram shows the wiring of this time clock and indicates that the time clock is manufactured by Tork and that the catalog number is 7300Z. The time clock is fed by four No. 12 AWG conductors in ¾-inch conduit from panel A.

Fig. 4-8 contains an electrical-symbol list for the project; a panelboard schedule, which gives the loads connected to each circuit breaker; and a power-wiring layout indicated as "Outlet Plan."

From the symbol list, it can be seen that the floor plan shows 31 duplex receptacles mounted 18 inches from the finished floor to the center of the box, 2 duplex receptacles with weatherproof covers and 6 floor-mounted receptacles. The drawings show all of these receptacles as fed by branch circuits in a manner similar to the lighting fixtures, except that broken lines are used to indicate the raceways to the receptacles rather than solid lines, which are used for the lighting fixtures. A broken circuit line (as indicated in the symbol list) indicates that the circuit raceway is to be installed concealed in the floor (concrete slab) or in the ceiling below.

The drawings also show that six wall-mounted telephone outlets are to be installed as well as three floor-mounted telephone outlets. Five computer outlets are also shown with empty ¾-inch conduit run from each outlet to the computer control panel.

The remaining power outlets are indicated by a junction-box symbol, which means that the circuits are connected directly to the various pieces of equipment. For example, the water heater (broken circle with "W.H.") is fed by circuit No. 8 containing two No. 10 AWG conductors and is connected to a two-pole 30-ampere circuit breaker in panel A. The HVAC (heating, ventilating, and air-conditioning) unit is fed by four 3-phase circuits; all contain three No. 6 AWG conductors and are connected to three-pole circuit breakers in panel A, rated at 60 amperes each. The pad-mounted condensing unit has a nonfusible weatherproof disconnect mounted on the unit and is fed with a circuit containing three No. 6 AWG conductors.

SUMMARY

The four basic types of building drawings found in a set of electrical drawings are: plans, elevations, sections, and details.

An orthographic-projection drawing is the most frequently used type of drawing for electrical systems in building construction.

The electrical floor plan of a building shows all outside walls, interior partitions, windows, doors, etc., along with the location of all electrical outlets, panelboards, branch circuits, feeders, service wire and equipment, and the other details necessary for making a correct installation.

The lighting layout for a building will usually be laid out on a separate floor plan from the power wiring.

Electrical symbols are used almost exclusively in showing the location of electrical outlets on building floor plans.

ASSIGNMENT 4

The following questions should be answered by filling in the blanks.

1. The panelboard schedule (Fig. 4-8) indicates that circuit No. 2 feeds receptacles and has a total connected load of _____ watts.

2. The number of computer outlets shown in Fig. 4-8 is _____.

3. Circuit A-1 (Fig. 4-8) feeds a three-phase _____ and contains three No. _____ conductors.

4. A _____ on the drawing in Fig. 4-8 indicates that the conduit feeding the computer outlets is _____ inch and is installed under _____.

5. There are four lighting fixtures shown at the entrance to the building in Fig. 4-7; two are type 6 and two are type 7. The circuit feeding these fixtures connects to panel _____.

6. How are the type-7 fixtures mounted (Fig. 4-7)?_____.

7. The type-7 lighting fixture (Fig. 4-7) is manufactured by _____, and the catalog number is _____.

8. The type-5 fixture (Fig. 4-7) contains _____, _____-watt F lamp.

9. The two type-4 lighting fixtures (Fig. 4-7) are _____ mounted.

10. How many type-2 fixtures are there? _____.

5

Sectional Views and Electrical Details

A section of any object, such as building, panelboard, and so forth, is what could be seen if the object was sliced or sawed into two parts at the point where the section was taken. For example, if we wished to see how a golf ball is constructed, we could place a golf ball in a vise and saw it in half with a hack saw; then we could easily see how the golf ball was constructed, or we would at least have a view of the internal construction. This typifies the need for sectional views in building construction drawings.

SECTIONING

Sometimes the construction of a building is difficult to show with the regular projection views we studied previously. If too many broken lines are needed to show hidden objects in the building, the drawings become confusing and difficult to read. Therefore, building sections are shown to clarify the construction. To better understand a building section, imagine that the building has been cut into sections as if with a saw. The floor plan of the building in Fig. 5-1 shows a sectional cut at point A-A. This cut is then shown in Fig. 5-2.

Let's assume that an electrician wishes to know the construction of the building walls in Fig. 5-1 in order to plan the layout of the outlet boxes which must be recessed in the wall. In order to see the cross section of the outside building wall, imagine a vertical cut through the outside wall on line A-A. The arrows indicate the direction from which the section is viewed. If the portion of the wall that is nearest the viewer were taken away, the remaining portion would be seen from the direction indicated by the arrows and would appear as shown in Fig. 5-2. The section itself is also identified by the title "Section A-A."

Now, the electrician can clearly see that the building wall is constructed of 8-inch thick masonry block. The electrician then knows that he must install his conduit carefully and work very closely with the masons constructing the block wall or expensive cutting and channeling will be necessary.

If underground conduit needs to enter the building through the footing, such as conduit for an underground service entrance, a sleeve could be inserted prior to pouring the concrete footings. Sleeves are short pieces of pipe placed in concrete or masonry to provide an opening for conduit or similar items that must be installed in the space at a later date. The opening avoids expensive cutting during construction, and it allows for conduit expansion and replacement.

A *chase* is another means of concealing conduit, surface ducts, etc., in masonry walls. The chase may be recessed as shown in Fig. 5-3, or the finished wall may have to be made thicker to enclose larger chase as shown in Fig. 5-4.

In dealing with sections, one must use a considerable amount of visualization. Some sections are very easy to visualize, while some are extremely difficult. There are no rules for determining what a section will look like. For example, a piece of rigid conduit, cut vertically, as shown in Fig. 5-4A, will have the shape of a rectangle; cut horizontally, as shown in Fig. 5-5B, it will be a circle; cut on the slant, as shown in Fig. 5-5C, it will be an ellipse.

Fig. 5-6 shows a pictorial (isometric) drawing of a building that will produce section A-A if cut vertically through the width (Fig. 5-7). If it is cut vertically through the length, it will produce section B-B, and if cut horizontally, it will produce section C-C. The cutting-plane line (Fig. 5-8) is used to show where the cutting plane is assumed to pass through an object.

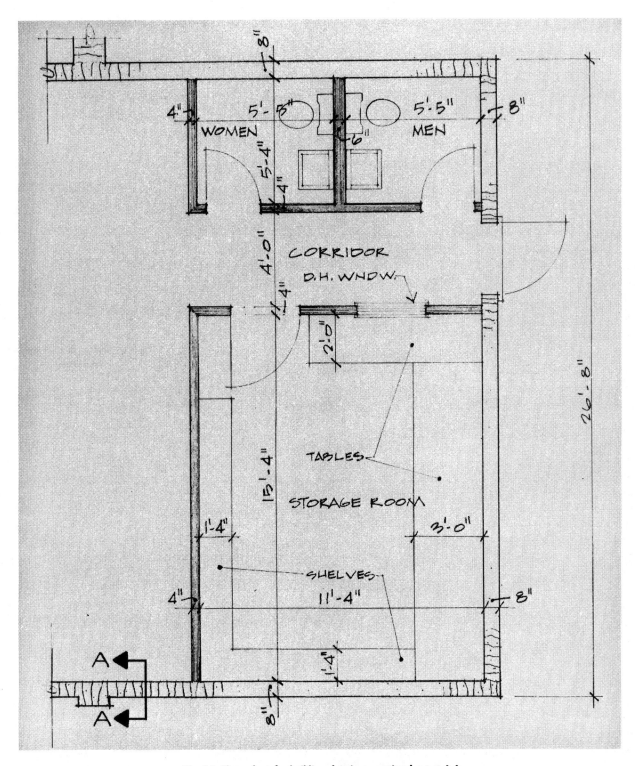

Fig. 5-1. Floor plan of a building showing a sectional cut at A-A.

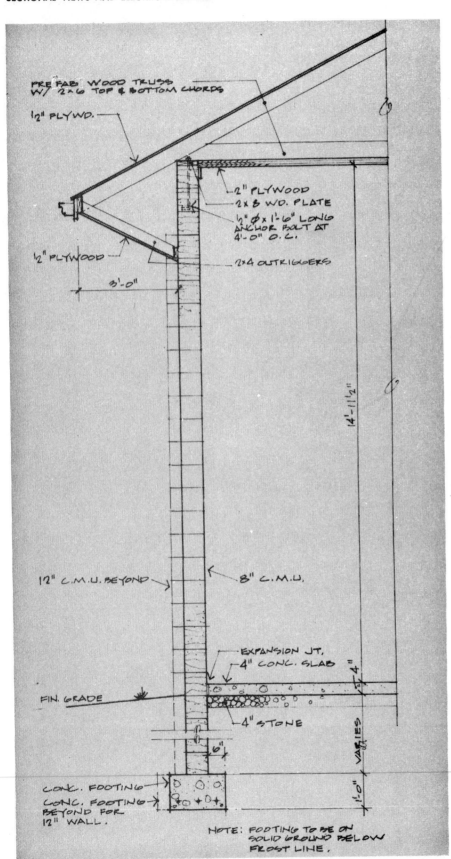

PREFAB. WOOD TRUSS
W/ 2×6 TOP & BOTTOM CHORDS

½" PLYWD.

2" PLYWOOD
2×8 WD. PLATE
½"∅ × 1'-6" LONG
ANCHOR BOLT AT
4'-0" O.C.

2×4 OUTRIGGERS

½" PLYWOOD

3'-0"

14'-11½"

12" C.M.U. BEYOND

8" C.M.U.

EXPANSION JT.
4" CONC. SLAB

4"

FIN. GRADE

4" STONE

6"

1'-0" VARIES

CONC. FOOTING
CONC. FOOTING
BEYOND FOR
12" WALL.

NOTE: FOOTING TO BE ON
SOLID GROUND BELOW
FROST LINE.

Fig. 5-2. Wall section of the cut at
A-A in Fig. 5-1.

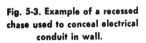

Fig. 5-3. Example of a recessed chase used to conceal electrical conduit in wall.

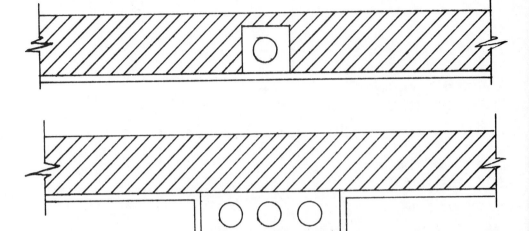

Fig. 5-4. An example of built-out chase on a solid wall.

Arrowheads on the ends of the cutting-plane line show the direction in which the section is viewed. Letters, such as *A-A, B-B,* etc., are normally used with cutting-plane lines to identify the cutting plane and the corresponding sectional views.

(A) Vertically.

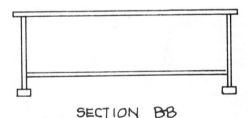

(B) Horizontally.   (C) On a diagonal.

Fig. 5-5. Electrical conduit cuts.

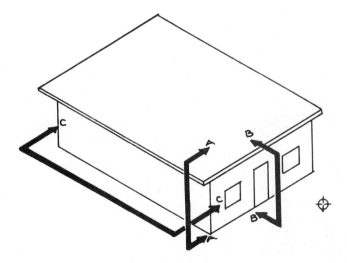

Fig. 5-6. Isometric drawing of a building showing sectional lines.

The views shown in Figs. 5-9 through 5-12 are typical sectional views of items used in electrical systems. They should be carefully studied and all details noted.

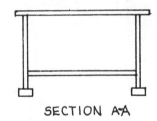

SECTION A-A

SECTION B-B

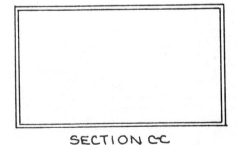

SECTION C-C

Fig. 5-7. Sections A-A, B-B, and C-C of the building in Fig. 5-6.

Fig. 5-8. Cutting-plane line.

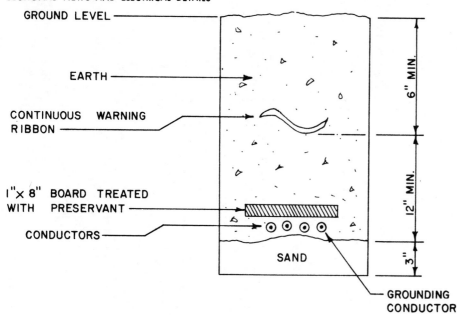

GROUND LEVEL

EARTH

CONTINUOUS WARNING RIBBON

1" × 8" BOARD TREATED WITH PRESERVANT

CONDUCTORS

6" MIN.

12" MIN.

3"

SAND

GROUNDING CONDUCTOR

Fig. 5-9. Example of a sectional cut through a trench in order to show details of buried wire.

The student should also practice drawing sections of different objects freehand in order to better understand the basic principles.

ELECTRICAL DETAILS

A detail drawing is a drawing of a separate item or portion of an electrical system, giving a complete and

exact description of its use and all the details needed to show the workman exactly what is required for its installation. The floor plan in Fig. 5-13 is a good example of an electrical drawing where an extra, detailed drawing is desirable.

It is obvious that the area contains lighting outlets evenly spaced in the ceiling and that the circuits feeding them are to be concealed. However, these outlets

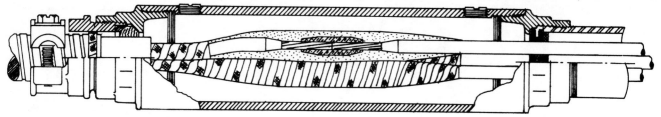

Fig. 5-10. Section through a cable splice.

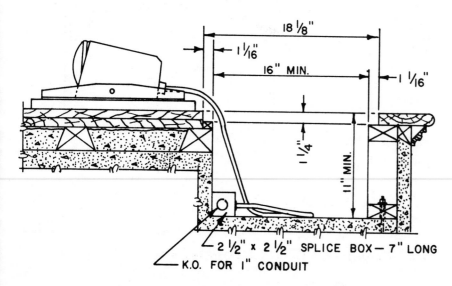

18 1/8"

1 1/16"

16" MIN.

1 1/16"

1 1/4"

11" MIN.

2 1/2" × 2 1/2" SPLICE BOX — 7" LONG

K.O. FOR 1" CONDUIT

Fig. 5-11. Sectional cut through a theatrical stage in order to show mounting details of a foldaway lighting fixture.

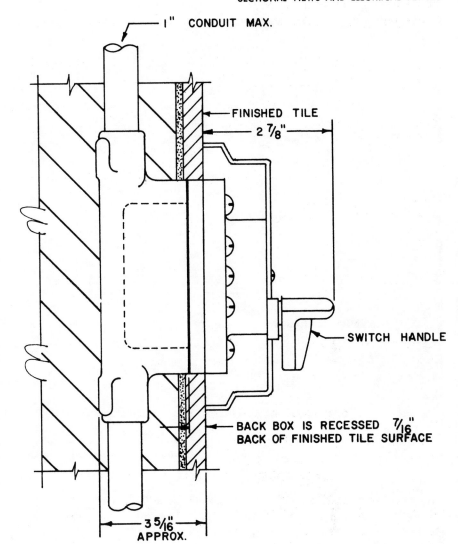

Fig. 5-12. Section through wall revealing installation details of an explosion-proof switch.

1" CONDUIT MAX.

FINISHED TILE

2 $\frac{7}{8}$"

SWITCH HANDLE

BACK BOX IS RECESSED $\frac{7}{16}$" BACK OF FINISHED TILE SURFACE

3 $\frac{5}{16}$" APPROX.

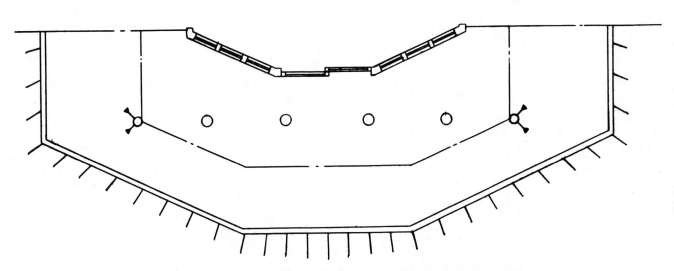

Fig. 5-13. Floor plan of a building utilizing concealed outlet boxes in a solid wood deck.

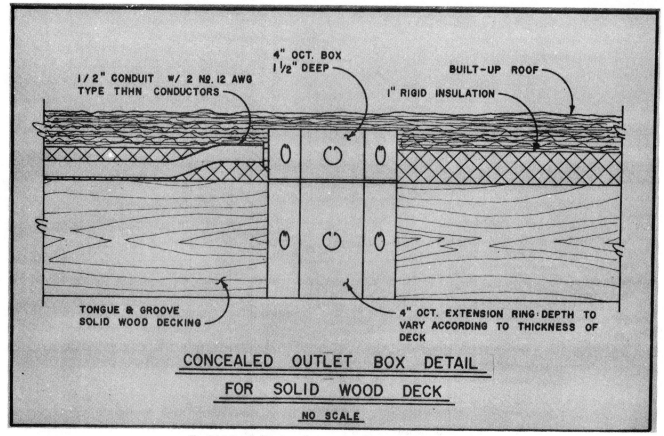

CONCEALED OUTLET BOX DETAIL

FOR SOLID WOOD DECK

NO SCALE

Fig. 5-14. Detail showing solution to installing concealed outlet box.

are located on a *solid* wood deck, and it would be difficult indeed to conceal conduit in such a deck. The detail, however, in Fig. 5-14 gives a simple solution to the problem and leaves little doubt as to exactly what is required.

A set of electrical drawings will sometimes require large-scale drawings of certain areas that are not indicated with sufficient clarity on the small-scale drawings. For example, the floor plan of the electrical room in Fig. 5-15 was originally drawn to a scale of $\frac{1}{16}'' = 1'0''$. At this scale, the equipment and related wiring is quite difficult to read, but a larger-scale drawing (Fig. 5-16) is much clearer.

The electrical details in Figs. 5-17 through 5-21 are typical of those encountered on electrical construction drawings. Some were taken directly from actual working drawings, while some were made available from electrical equipment manufacturers. All should be studied, and every detail carefully noted.

The tv-outlet detail in Fig. 5-17 shows a 4-inch square box with a single-gang plaster ring and tv-outlet plate mounted on a piece of ¾-inch conduit that terminates under the floor in the crawl space of the building. The distance from the center of the box to the finished floor is 16 inches. After the building is roughed-in, the tv-cable company will install their coaxial cable in the crawl space and then fish the wire up to each tv-outlet box.

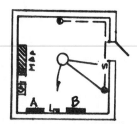

Fig. 5-15. Floor plan of an electrical room drawn to a scale of 1/16″ = 1′-0″.

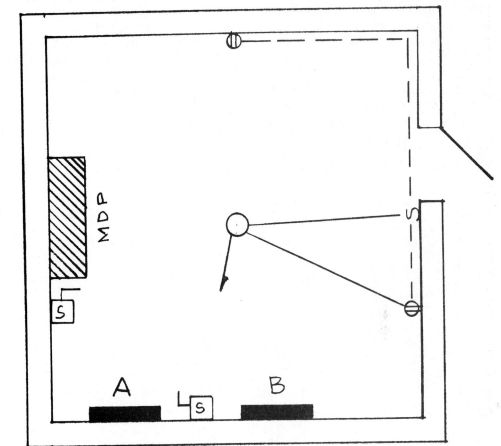

Fig. 5-16. A larger-scale drawing of Fig. 5-15.

Fig. 5-18 shows details of the installation of telephone-outlet boxes in a building. Note that the telephone-outlet box is mounted 18 inches above the floor to the CL of the box and that 1-inch conduit is run inside the wall up above the finished lay-in tile ceiling. The telephone installers will then run their cable in the space above the finished ceiling and fish their telephone cables down the conduit to each outlet box.

The spire-lighting detail in Fig. 5-19 shows a lighting fixture mounted on an outlet box and supported by a piece of 1¼-inch conduit. This outlet box and conduit is for support only, since the electrical current is pro-

4" SQ. BOX / SINGLE GANG PLASTER RING AND TV OUTLET PLATE

¾" CONDUIT

FLOOR

CRAWL SPACE

16"

BUSHING

Fig. 5-17. Tv-outlet detail.

vided by a flexible cord plugged into an outlet. With this arrangement, the narrow-beam spotlight will shine up into the spire and giving it the effect of glowing at night.

Fig. 5-20 shows the details of connection and installation of a residential post lamp. The detail shows that the post is to be installed two feet deep in the ground, the earth is to be tamped around it, and the portion of the post above the ground is to be four feet. A two-wire UF cable with ground is to be buried directly in the group up to the source of power, up the post (inside), and the cable is then connected to the lighting fixture.

Fig. 5-21 shows an LP/gas-tank connection detail

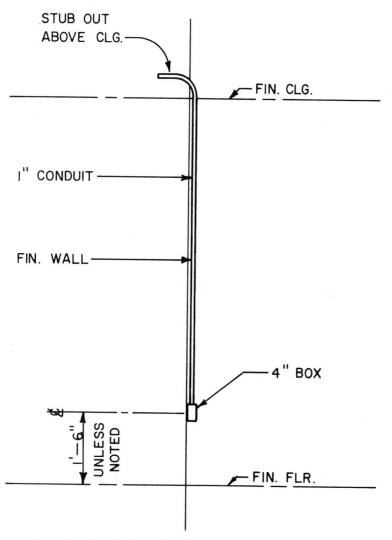

Fig. 5-18. Telephone-outlet connection.

that was used on an electrical drawing for a firehouse that needed a standby emergency generator for certain electrical loads within the building. All necessary details of construction are shown in this detail; that is, the thickness of sand around the tank, the capacity of the tank, the location of the shutoff valve, etc.

SUMMARY

A sectional view is one in which a portion of the object is assumed to be removed to reveal the interior details. In dealing with sections, one must use a considerable amount of visualization.

An electrical detail drawing is a drawing of a single item or a portion of an electrical system; it gives all the necessary details and a complete description of its use in order to show the workman exactly what is required for installation.

A set of electrical drawings will sometimes require large-scale drawings of certain items that are not indicated with sufficient clarity on the small-scale drawings to simplify reading.

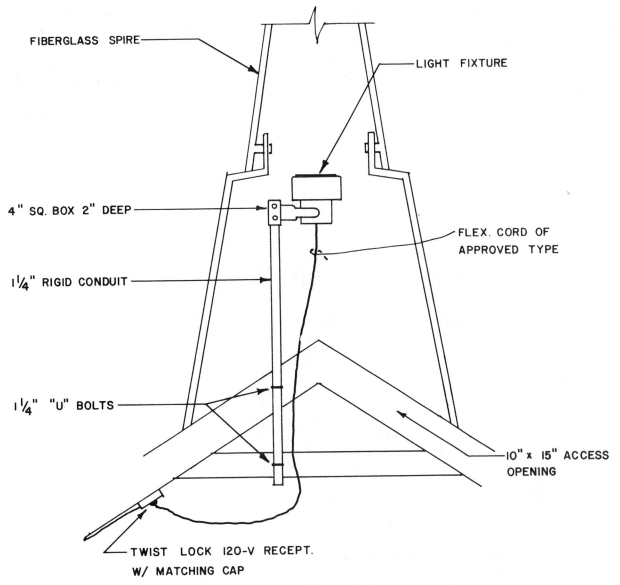

FIBERGLASS SPIRE

LIGHT FIXTURE

4" SQ. BOX 2" DEEP

FLEX. CORD OF
APPROVED TYPE

1¼" RIGID CONDUIT

1¼" "U" BOLTS

10" x 15" ACCESS
OPENING

TWIST LOCK 120-V RECEPT.
W/ MATCHING CAP

Fig. 5-19. Spire-lighting detail.

ASSIGNMENT 5

Answer the follwing questions by filling in the blanks.

1. A section of any object is what could be seen if the object was _____ or _____ into two parts at the point where the section was taken.

2. Building sections are shown to _____ the construction.

3. How thick is the masonry block in Fig. 5-2? _____.

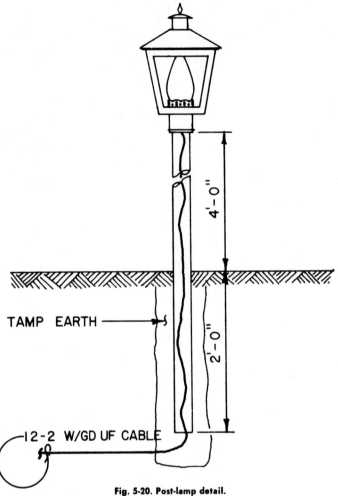

4'-0"

TAMP EARTH

2'-0"

12-2 W/GD UF CABLE

Fig. 5-20. Post-lamp detail.

4. If conduit is to be installed in a solid masonry wall, a _____ or a _____ may be used to simplify the installation.

5. In dealing with building sections, a person must use a considerable amount of _____.

6. A drawing showing a separate item or portion of an electrical system and giving complete and exact descriptions of its use and installation is called a _____ drawing.

7. What is installed on the end of the ¾-inch conduit in Fig. 5-17? _____.

8. What size conduit is used in the detail in Fig. 5-18? _____.

9. What size "U" bolts are used to support the 1¼-inch conduit in Fig. 5-19? _____.

10. What is the kilowatt rating of the electric generator in Fig. 5-21? _____.

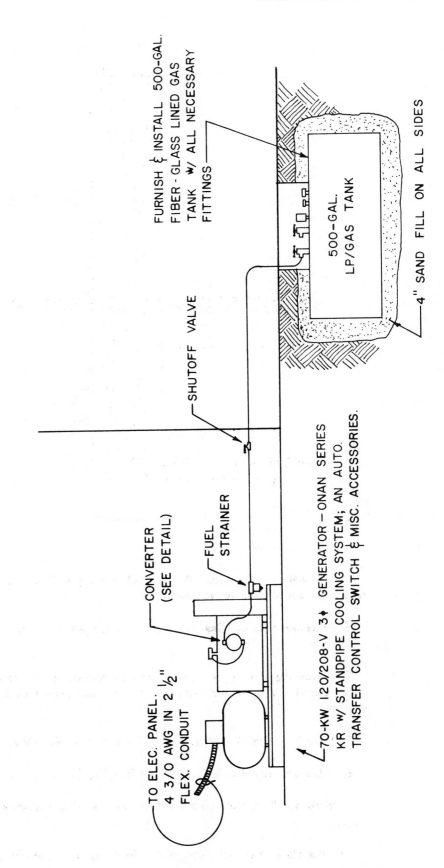

Fig. 5-21.
Detail of LP/gas-tank connection.

FURNISH ϟ INSTALL 500-GAL. FIBER-GLASS LINED GAS TANK W/ ALL NECESSARY FITTINGS

500-GAL. LP/GAS TANK

4" SAND FILL ON ALL SIDES

SHUTOFF VALVE

CONVERTER (SEE DETAIL)

FUEL STRAINER

TO ELEC. PANEL. 4 3/0 AWG IN 2 1/2" FLEX. CONDUIT

70-KW 120/208-V 3ϕ GENERATOR — ONAN SERIES KR w/ STANDPIPE COOLING SYSTEM; AN AUTO. TRANSFER CONTROL SWITCH ϟ MISC. ACCESSORIES.

6

Electrical Wiring Diagrams

A large part of all electrical drawing deals with circuits. Therefore, it is important that those required to interpret electrical drawings have a thorough understanding of the more common circuits used in electrical systems for building construction. These circuits are usually shown by one or all of the following methods.

1. Diagrammatic plan views showing individual building-circuit layouts.
2. Complete schematic diagrams showing all details of connection and every wire in the circuit.
3. One-line diagrams.
4. Power-riser diagrams.

The first method of showing electrical circuits was covered in Chapter 4, so no further explanation of diagrammatic plan views will be given in this chapter.

Complete schematic wiring diagrams are normally used only in highly unique and complicated electrical systems, such as control circuits. Components are represented by symbols, and every wire is either shown by itself or included in an assembly of several wires which appear as one line on the drawing. Each wire should be numbered when it enters an assembly and should keep the same number when it comes out again to be connected to some electrical component in the system. Fig. 6-1 shows a complete schematic wiring diagram for a three-phase, ac magnetic non-reversing motor starter.

Note that this diagram shows the various devices in symbol form and indicates the actual connections of all wires between the devices. The three-wire supply lines are indicated by L_1, L_2, and L_3; the motor terminals of motor M are indicated by T_1, T_2, and T_3. Each line has a thermal overload-protection device (OL) connected

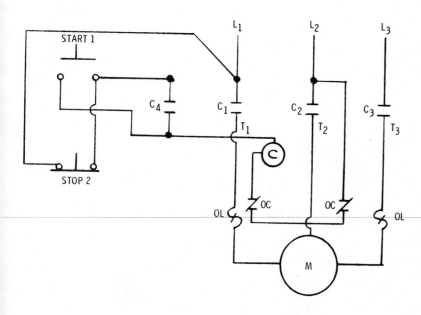

Fig. 6-1. Wiring diagram for a three-phase, ac magnetic nonreversing motor starter.

in series with normally open line contactors C_1, C_2, and C_3, which are controlled by the magnetic starter coil, C. Each contactor has a pair of contacts that close or open during operation. The control station, consisting of start push button 1 and stop push button 2, is connected across lines L_1 and L_2. An auxiliary contactor (C_4) is connected in series with the stop push button and in parallel with the start push button. The control circuit also has normally closed overload contactors (OC) connected in series with the magnetic starter coil (C).

Any number of additional push-button stations may be added to this control circuit similarly to the way three- and four-way switches are added to control a lighting circuit. In adding push-button stations, the stop buttons are always connected in series and the start buttons are always connected in parallel. Fig. 6-2 shows the same motor starter circuit in Fig. 6-1, but this time it is controlled by two sets of start-stop buttons.

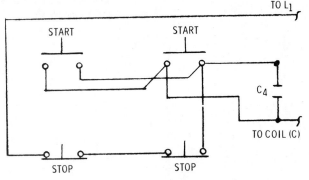

Fig. 6-2. Same wiring diagram in Fig. 6-1 being controlled by two sets of start-stop buttons.

SINGLE-LINE DIAGRAMS

Fig. 6-3 gives a list of electrical wiring symbols commonly used for single-line schematic diagrams. Fig. 6-4 shows a typical single-line diagram of an industrial power-distribution system. In analyzing this diagram, refer to the symbol list often.

The utility company will bring their lines (from 22.9 to 138 kV) to a substation outside the plant building. Here, air switches, lightning arresters, single-throw switches, and an oil circuit breaker are provided. This substation also reduces the primary voltage to 4160 volts by means of transformers. Again, lightning arresters and various disconnecting means are shown.

Next, the 4160-volt service enters the building and is metered. Air circuit breakers are shown with disconnecting means on each side. From this point, the 4160-volt services are routed to various locations within

POWER EQUIPMENT

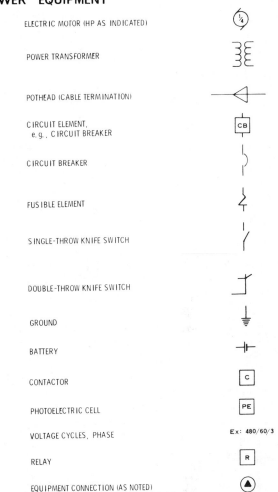

ELECTRIC MOTOR (HP AS INDICATED)	
POWER TRANSFORMER	
POTHEAD (CABLE TERMINATION)	
CIRCUIT ELEMENT, e.g., CIRCUIT BREAKER	CB
CIRCUIT BREAKER	
FUSIBLE ELEMENT	
SINGLE-THROW KNIFE SWITCH	
DOUBLE-THROW KNIFE SWITCH	
GROUND	
BATTERY	
CONTACTOR	C
PHOTOELECTRIC CELL	PE
VOLTAGE CYCLES, PHASE	Ex: 480/60/3
RELAY	R
EQUIPMENT CONNECTION (AS NOTED)	

Fig. 6-3. Electrical wiring symbols commonly used for single-line diagrams.

the plant and are then subdivided into feeders for several different departments. Again, air circuit breakers and disconnecting means are provided for each feeder.

The "Aux Feeder" connects to another transformer where the voltage is reduced to 480 volts for feeding a motor control center. The "Dept. Feeder" immediately to the right is also reduced to 480 volts to feed a motor control center. The next department feeder is also used to feed 480-volt motor control centers.

Most of the other feeders are also reduced to 480 volts to feed motor control centers. However, a few feeders are used to supply energy to motors rated at the full voltage. There is also a "generator tie" circuit.

Notice that one of the feeders continues to another substation. Although not shown on this diagram, the voltage will probably be reduced at this point to 120/208 volts for use on 120-volt lighting and convenience outlets.

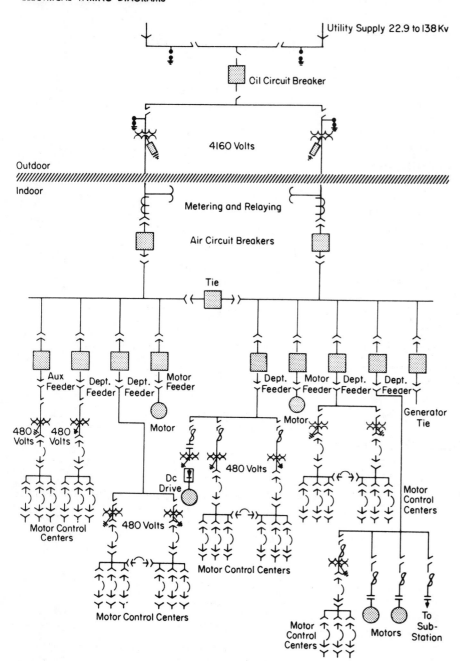

Utility Supply 22.9 to 138 Kv

Oil Circuit Breaker

4160 Volts

Outdoor

Indoor

Metering and Relaying

Air Circuit Breakers

Tie

Aux Feeder

Dept. Feeder

Dept. Feeder

Motor Feeder

Dept. Feeder

Motor Feeder

Dept. Feeder

Dept. Feeder

Generator Tie

Motor

Motor

480 Volts

480 Volts

Dc Drive

480 Volts

Motor Control Centers

Motor Control Centers

480 Volts

Motor Control Centers

Motor Control Centers

Motor Control Centers

Motors

To Sub-Station

Fig. 6-4. Example of a single-line diagram used in an industrial power distribution system.

Typical Power Distribution System for Industrial Plants

Courtesy Westinghouse

RISER DIAGRAMS

Power-riser diagrams are probably the most frequently used type of diagrams on electrical working drawings for building construction. Such diagrams give a picture of what components are to be used and how they are to be connected in relation to one another. This type of diagram is easily understood and requires much less time to draw or interpret than the previously described types of diagrams. As an example, compare the power-riser diagram in Fig. 6-5 with the schematic diagram in Fig. 6-6. Both are diagrams of an identical electrical system, but it is easy to see that the drawing in Fig. 6-5 is greatly simplified.

The power-riser diagram in Fig. 6-7 was used on a working drawing for a printing-company building. The main building service was to be 3-phase, 4-wire, 120/208-volt connected. However, several old single-phase motors rated at 240 volts were to be used in some of the printing machines. Since running these motors on

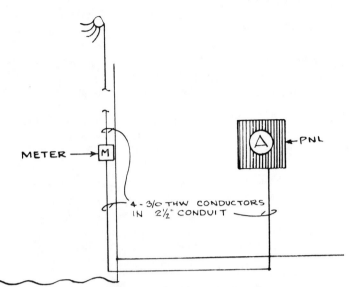

Fig. 6-5. Example of a typical power-riser diagram.

two phases of the 208-volt service would greatly shorten the life of each 240-volt motor, it was decided to specify an additional panel (C) for use with the 240-volt equipment. This panel would connect to two booster transformers "Y-Δ," as shown in Fig. 6-8, in order to gain the additional voltage.

The power-riser diagram in Fig. 6-9 combines a single-line diagram with a conventional "block" riser diagram to convey the information. It is clear that the pad-mounted transformer, current transformers (CT), and watt-hour meter are furnished by the utility company

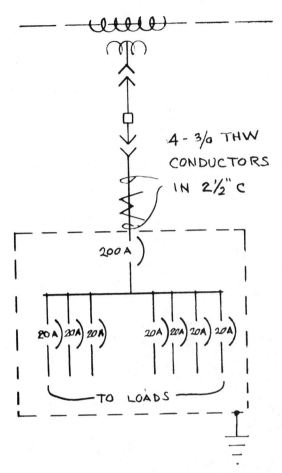

Fig. 6-6. Example of a one-line wiring diagram of the same service as shown in Fig. 6-5.

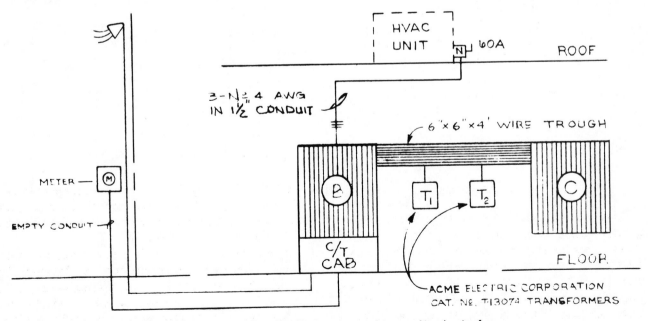

Fig. 6-7. Example of another power-riser diagram used on a working drawing for a printing-company building.

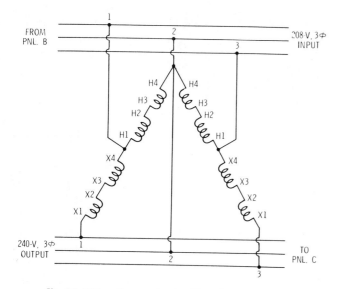

Fig. 6-8. Wiring diagram showing "boosting" transformers.

(VEPCO); the feeder conductors from the transformer to the main distribution panel are also furnished by the

utility company. However, the electrical contractor is required to furnish six 4-inch conduits for the utility company's conductors.

Everything else in this diagram is to be furnished and installed by the electrical contractor. From some of the details shown, we learn the following information:

1. The main distribution panel bus is rated for 2500 amperes, and the service characteristics are 3-phase, 4-wire, "Y" connected; that is, 120/208 volts.

2. A bolt-lock switch in the CT compartment is fused with 1600-ampere current-limiting fuses.

3. An emergency panel is fed by four No. 2 AWG conductors in 1¼-inch conduit and is connected ahead of the main switch so that if the main switch is opened, the emergency panel will still be energized.

This riser diagram does not show the overcurrent devices in the main distribution panel for the 11 feeders, but a schedule shown elsewhere on the drawings gives

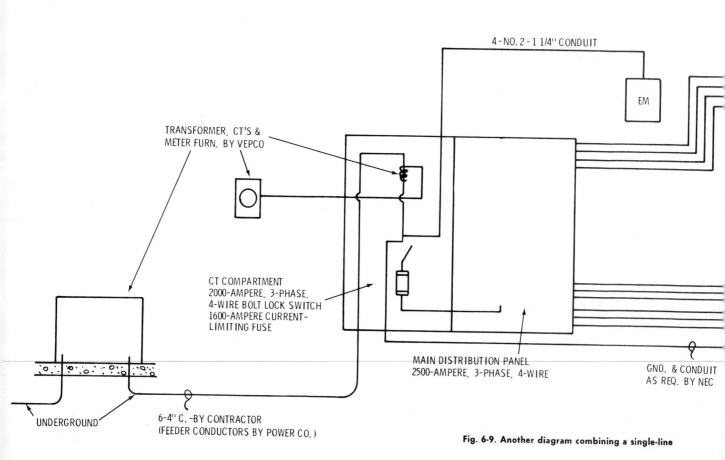

Fig. 6-9. Another diagram combining a single-line

all necessary details (Fig. 6-10). The number and size of all conductors feeding the subpanels on the first and second floor are also indicated in the panelboard schedule.

Fig. 6-11 shows a typical telephone-conduit riser diagram. While all of the outlets and telephone cabinets are shown on the project floor plans, no conduit was shown. Therefore, the main purpose of this riser diagram is to show the size of conduit required to each of the various outlets. The written specifications also state that the electrical contractor shall provide only the conduit and a galvanized pull-wire. The telephone company will install the telephone cables and make the necessary connections.

Since this project is a school, a clock and class-bell system is specified. Again, the location of the various clocks and bells are shown on the floor plans, but a riser diagram (Fig. 6-12) is also required to clearly indicate the connections of each.

A water-sprinkler system is provided in this building, and a wiring diagram (Fig. 6-13) is used to sound the fire-alarm system (Fig. 6-14) if any part of the sprinkler system in put into use. The fire-alarm system is activated by a flow switch; that is, the flow of any water in the sprinkler system closes a contact which energizes the fire-alarm system.

The fire-alarm riser diagram in Fig. 6-14 shows the main fire-alarm-system control cabinet which is fed by circuit No. 1 in panel EM. A ¾-inch conduit is also provided for connection to the existing fire-alarm system in another building, and the size and number of conductors are to be those recommended by the equipment manufacturer.

Single lines run from the main control cabinet to various alarm bells and striking stations. Again, a note on the drawings states: "Quantity and size of conductors as recommended by manufacturer of fire-alarm system." This diagram, then, is not complete and requires the contractor or those installing the system to obtain the necessary data from the manufacturers in order to install the system properly. However, it is not unusual to find cases like this on many electrical drawings. One

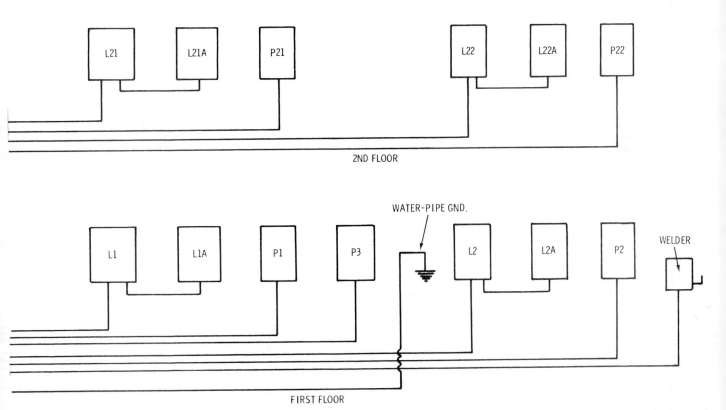

diagram with a conventional "block" riser diagram.

PANEL BOARD SCHEDULE

120/208V   PANEL -"A"   3Ø 4WIRES

SQUARE "D" TYPE NQOB W/200 A MAIN LUGS ONLY

CKT No.	CIRCUIT BREAKER			WIRE SIZE	CONNECTED LOAD IN KW			ITEMS FED OR REMARKS
	POLE	TRIP	FRAME		A Ø	B Ø	C Ø	
1	1	20	70	Nº12	1400			LIGHTS
2					1400			LIGHTS
3						1400		LIGHTS
4						1400		LIGHTS
5							1400	LIGHTS
6							1400	LIGHTS
7					1200			LYTESPAN TRACK
8					1200			LYTESPAN TRACK
9						1200		LYTESPAN TRACK
10						1200		LYTESPAN TRACK
11							500	SPARE
12							1200	RECEPTS.
13					1200			
14					1200			
15						1200		
16						1200		
17							1200	
18							1200	
19	3	30	70	Nº10	2340			ROOFTOP UNIT Nº2
20	3	80	100	Nº4	7188			ROOFTOP UNIT Nº1
21	___	___	___	___		2340		_____
22	___	___	___	___		7188		_____
23	___	___	___	___			2340	_____
24	___	___	___	___			7188	_____
25	1	20	70	___	___	___	___	SPARE
26								
27								
28								
29								
30								
31	___	___	___	___	___	___	___	PROVISION ONLY
32	___	___	___	___	___	___	___	PROVISION ONLY
TOTAL CONNECTED LOAD					17,128	17,128	16,428	

Fig. 6-10. Panelboard schedule giving additional data necessary for use with the riser diagram in Fig. 6-9.

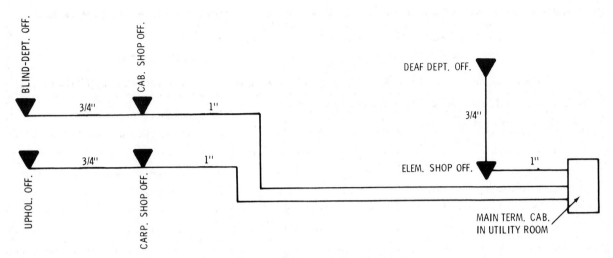

Fig. 6-11. An example of a typical telephone-conduit riser diagram.

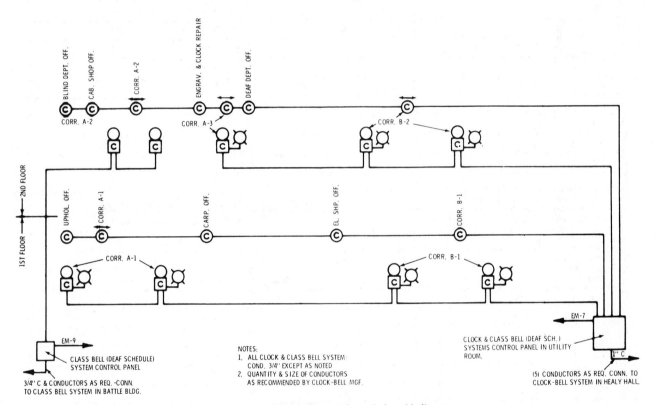

Fig. 6-12. Example of a riser diagram for a clock and bell system.

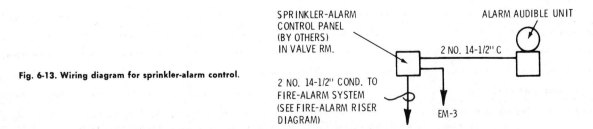

Fig. 6-13. Wiring diagram for sprinkler-alarm control.

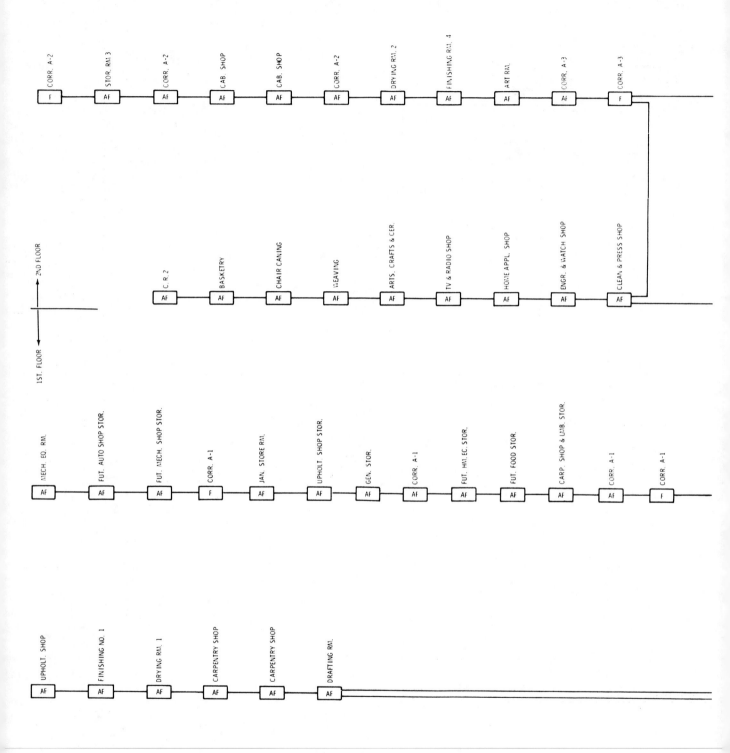

Fig. 6-14. Example of a

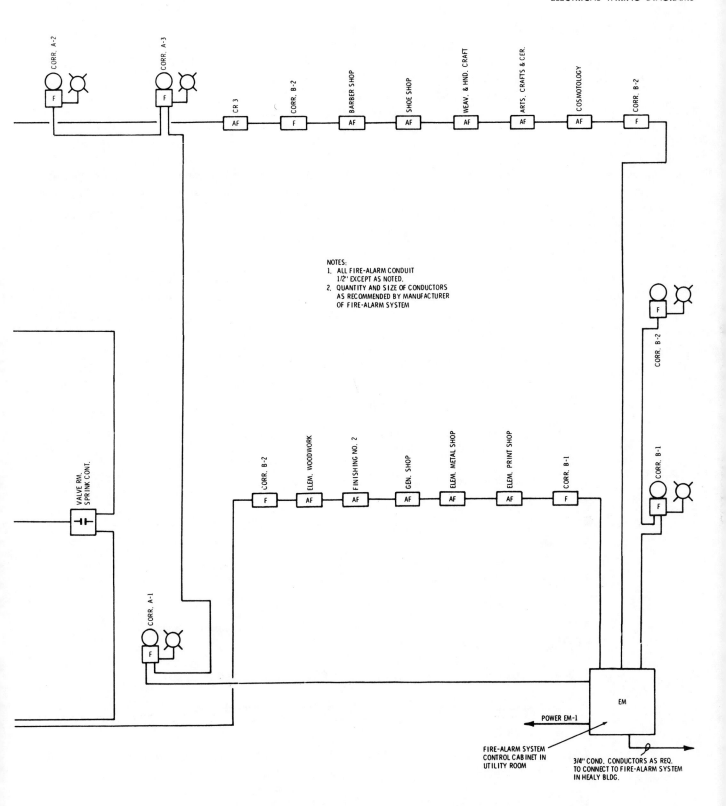

NOTES:
1. ALL FIRE-ALARM CONDUIT 1/2" EXCEPT AS NOTED.
2. QUANTITY AND SIZE OF CONDUCTORS AS RECOMMENDED BY MANUFACTURER OF FIRE-ALARM SYSTEM

CORR. A-2

CORR. A-3

CR 3

CORR. B-2

BARBER SHOP

SHOE SHOP

WEAV. & HND. CRAFT

ARTS. CRAFTS & CER.

COSMOTOLOGY

CORR. B-2

CORR. B-2

CORR. B-1

VALVE RM. SPRINK CONT.

CORR. B-2

ELEM. WOODWORK

FINISHING NO. 2

GEN. SHOP

ELEM. METAL SHOP

ELEM. PRINT SHOP

CORR. B-1

CORR. A-1

EM

POWER EM-1

FIRE-ALARM SYSTEM CONTROL CABINET IN UTILITY ROOM

3/4" COND. CONDUCTORS AS REQ. TO CONNECT TO FIRE-ALARM SYSTEM IN HEALY BLDG.

fire-alarm riser diagram.

reason may be that the existing system is obsolete and a suitable new system has to be engineered, which takes more time than can be had before the building construction is to begin. The lack of such data will invari-ably cause the bid to be higher than it normally would be if all the necessary information were provided.

Electrical circuits are usually shown by one or all of the following methods:

SUMMARY

Electrical circuits are usually shown by one or all of the following methods:

1. Plan views drawn to scale showing the location of all outlets and their related wiring.
2. Complete schematic diagrams showing all details of the various components and every wire in the circuit connecting them.
3. Single-line diagrams which are a simplified version of complete schematic diagrams.
4. Simplified block diagrams, often referred to as power-riser diagrams or any other type of riser diagram, such as telephone, fire alarm, etc.

ASSIGNMENT 6

Answer the following questions by filling in the blank spaces.

1. The motor-control diagram in Fig. 6-1 has _____ normally open line contactor(s) and _____ auxiliary contactor(s).

2. The three motor terminals in Fig. 6-1 are designated _____, _____, and _____.

3. The line voltage after the oil circuit breakers and transformers in Fig. 6-4 is _____ volts.

4. From the symbol list in Fig. 6-3, the symbol $\exists\xi$ indicates a _____ transformer.

5. What is the ampere rating of the bolt-lock switch in the CT compartment in Fig. 6-9? _____ amperes.

6. The two sizes of conduit indicated in Fig. 6-11 are _____ and _____.

7. The ground wire in Fig. 6-9 is indicated as being connected to a _____ ground.

8. The wire size specified in Fig. 6-13 is two No. _____ AWG and is installed in _____-inch conduit.

9. On what floor is panel "P3" shown in Fig. 6-9? _____.

10. The grounding wire and conduit in Fig. 6-9 are to be sized and installed according to _____.

Electrical Schedules

A schedule, as related to electrical drawings, is a systematic method of presenting notes or lists of equipment on a drawing in tabular form. When properly organized and thoroughly understood, schedules are not only powerful timesaving methods for the draftsman and engineer, but also save the specification writer and the workmen on the job much valuable time.

For example, the lighting-fixture schedule shown in Fig. 7-1 lists the fixture type and identifies each fixture type on the drawing by number. The manufacturer and catalog number of each type are given along with the number, size, and type of lamp for each. The "Volts" and

"Mounting" columns follow, and a column is left for remarks. This latter column may give such information as the mounting height above the finished floor, in the case of a wall-mounted lighting fixture, or any other data required for the proper installation of the fixtures.

Sometimes all of the same information can be found in the specifications of the project, but combing through page after page of written specifications can be time consuming and workmen do not always have access to the specifications while working, whereas they usually do have access to the working drawings. Therefore, the schedule is an excellent means of providing essential in-

LIGHTING-FIXTURE SCHEDULE						
FIXT. TYPE	MANUFACTURER'S DESCRIPTION	LAMPS NO.	TYPE	VOLTS	MOUNTING	REMARKS

Fig. 7-1. Typical lighting-fixture schedule.

CONNECTED-LOAD SCHEDULE	
TYPE OF LOAD	TOTAL IN KVA
LIGHTING	14.9
RECEPTACLES & MISC.	8.2
AIR CONDITIONING	18.3
WATER HEATER	1.5

Fig. 7-2. Connected-load schedule.

formation in a clear and accurate manner, allowing the workmen to carry out their assignments in the least amount of time.

The following schedules are typical of those used on electrical drawings by consulting engineering firms. Each should be thoroughly studied.

CONNECTED-LOAD SCHEDULE

When the electrical layout for a building is completed, the utility company often requires a site plan of the building plot (see Chapter 8) and a breakdown of the total connected load in order to size the service conductors, transformers, etc. The electrical designer normally supplies this information by means of a form letter giving all necessary data. However, it is sometimes necessary to show this information on the working drawings, especially when government funds are used on the project in question. When this is required, the "Connected-Load Schedule" in Fig. 7-2 may be used.

The left-hand column is provided for a general description of the electrical load type; in this case it is broken down into lighting, receptacles and miscellaneous, air conditioning, and water heater. The right-hand column—"Total in kVA"—is the same as kW, or kilowatts. Therefore, we know the lighting load totals 14.9 kW; the receptacle and miscellaneous load totals 8.2 kW; the air conditioning load is 18.3 kW; and the water heater is 1.5 kW. When these are totaled, the inspector or utility company engineer will use these figures to check the service-entrance capacity, which is specified elsewhere on the drawings.

$$\begin{array}{r} 14.9 \\ 8.2 \\ 18.3 \\ 1.5 \\ \hline 42.9 \text{ kW} \end{array}$$

Therefore, the main service should be approximately:

$$\frac{42.9 \text{ (kW)} \times 1000 \text{ (to convert to watts)}}{208 \text{ (volts)} \times 3 \text{ (3-phase factor)}} = 68.75 \text{ A}$$

The utility company will probably use a demand factor to size their transformer, and the transformer size will probably be:

CONNECTED LOAD

BLDG.	LIGHTING KW	RECEPTACLES KW*	HEATERS KW	MOTORS HP	TOTAL KW
A	9.5	26	8	28	71.5
B	9.5	26	8	28	71.5
C	19	27.5	8	34	88.5

LARGEST SINGLE MOTOR — 2 HP.

*ALLOWANCE FOR MINIMUM USE OF PLUG-IN APPLIANCES BY TENANTS.

Fig. 7-3. Another type of connected-load schedule.

$$68.75\ (A) \times 0.65\ (\text{diversity factor}) = 44.68\ A$$

$$\frac{44.68\ (A) \times 208 \times 3\ (\text{volts, 3-phase})}{1000\ (\text{to convert to kVA})} = 27.89\ kVA$$

Thus, the utility company would probably install a transformer approximately 28 to 30 kVA in capacity.

This particular schedule may need to be modified for each individual job for such additional loads as electric heating, total motor horsepower, and the largest single motor. The "Connected-Load Schedule" in Fig. 7-3 was taken from an apartment project and shows these additional loads.

PANELBOARD SCHEDULES

The panelboard schedule in Fig. 7-4 is typical of the so-called short forms in that it provides sufficient data to identify the size and type of the panelboard but does not give detailed information concerning the individual circuits; this latter information must be given elsewhere on the drawings (usually in the plan views).

A practical application of this form is shown in Fig. 7-5. Here, the panelboard type is identified by the letter M. The location of this panelboard will be shown on the floor plan by an appropriate system and identified

PANELBOARD SCHEDULE

PANEL NO.	TYPE CABINET	PANEL MAINS			BRANCHES					ITEM FED OR REMARKS
		AMPS	VOLTS	PHASE	1P	2P	3P	PROT	FRAME	

Fig. 7-4. Panelboard schedule.

PANELBOARD SCHEDULE

PNL TYPE	TYPE CABINET	PANEL MAINS			BRANCHES					ITEMS FED
		AMPS	VOLTS	PHASE	1P	2P	3P	PROT	FRAME	
"M"	SURFACE	200A	120/240	1∅ 3W	15	—	—	20	70	LTS. & RECEPTS.
"ITE"	TYPE FEQ				—	4	—	20	70	HT. & COMP.
					—	1	—	30	70	W.H. & UNIT HTS.
					2	—	—	20	70	SPARES
					1	—	—	—	—	PROVISIONS ONLY

Fig. 7-5. Panelboard schedule with spaces filled in.

|							
LIGHTING PANEL A15*							
QO 125-A 120/240-V 1-PHASE 3-WIRE SN LUGS GROUND BAR							
CIR. NO.	WATTS	CIRCUIT BREAKER		CONDUCTOR			REMARKS
		NO. POLES	AMPS.	NO.	SIZE		
1	450	1	15	2	14		LIGHTING
2	200			2	14		RECEPTACLES
3							" —TELEPHONE
4	600		20	2	12		"
5	4 KW	2	30	2	10		WALL HEATER
6	4 KW	2	30	2	10		WALL HEATER
7							SPARE
8							"
* PANEL B15 IDENTICAL.							

Fig. 7-6. An example of another panelboard schedule.

as panel M. The schedule indicates that the panel cabinet is to be the surface-mounting type and is to contain a 200-ampere main circuit breaker. It also indicates that it is rated for 120/240 volts and will be fed with a single-phase (1 ϕ), 3-wire feeder. The panelboard is to be manufactured by ITE and is to be type FEQ.

The columns under the heading "Branches" give data pertaining to the overcurrent protection for the individual branch circuits. The schedule indicates that this panel will contain fifteen 1-pole circuit breakers with a "trip" rating of 20 amperes built on a 70-ampere frame; these circuit breakers provide overcurrent protection for lighting and receptacles as indicated in the "Items Fed" column.

The schedule calls for four 2-pole, 20-ampere breakers to feed heating units and a compressor. One 30-ampere, 2-pole breaker is provided for a circuit for a water heater and unit heater. The schedule also calls for two 20-ampere circuit breakers to be installed for future circuits and calls for space to be left for a third circuit breaker of any ampere rating (1-pole), also for a future circuit. From this, it is evident that the panelboard must be sufficiently large to accomodate 28 spaces.

The wire size is not indicated, but the electrician installing the system knows that a 20-ampere circuit requires No. 12 AWG wire and a 30-ampere circuit requires No. 10 AWG wire. However, the wire sizes are sometimes indicated on the floor-plan drawings when the outlets are circuited.

Fig. 7-6 shows another type of panelboard schedule used on an apartment project. This type has the circuits numbered, which means that the load should be balanced. It gives the total load on each circuit, the size and type of overcurrent protection, and the wire (conductor) size and number. The items fed are also listed in the "Remarks" column. While the manufacturers are not specified in this schedule, they are specified in the written specifications. Fig. 7-7 gives another type of panelboard schedule.

ELECTRIC-HEAT SCHEDULE

The electric-heat schedule in Fig. 7-8 is an excellent means of conveying necessary data for the installation of electric baseboard heaters. An identification mark is used with the location of the heaters on the floor plan. Then, adjacent to the mark on the schedule, manufacturers' data are entered and the physical dimensions, the rated voltage, the type of mounting, and remarks concerning the connection or mounting of these units are given.

KITCHEN-EQUIPMENT SCHEDULE

Fig. 7-9 shows a kitchen-equipment schedule used on the electrical working drawings for a commercial kitchen. The column designated "Equip. No." identifies the equipment on the floor plan; that is, the respective number inside the hexagon is placed adjacent to the piece of equipment on the floor with a circuit home run shown (Fig. 7-10); then the necessary data are entered in the schedule to provide further information for the

PANEL SCHEDULE

Panel No.		Designer		Checked		Date	
MAINS	Ø	W		Volts	Ampere		

Circuit No.	Switch or Breaker			SERVES	Connected KVA	Demand Factor	Demand KVA
	Pole	Frame	Trip or Fuse				
1							
2							
3							
4							
5							
6							
7							
8							
9							
10							
11							
12							
13							
14							
15							
16							
17							
18							
19							
20							
21							
22							
23							
24							
25							
26							
27							
28							
29							
30							
31							
32							
33							
34							
35							
36							
37							
38							
39							
40							
41							
42							
				TOTALS			

DEMAND I
DESIGN I   PERMISSIBLE Ed.   FEEDER SIZE   Ed/1000 A.F.
FEEDER LENGTH   AMP. FT.   M.   ACTUAL Ed.

Fig. 7-7. A panelboard schedule used by a consulting engineering firm.

ELECTRIC - HEAT SCHEDULE

HT'R TYPE	MANUFACTURER'S DESCRIPTION	DIMENSIONS	VOLTS	MOUNTING	REMARKS

Fig. 7-8. Electric-heat schedule.

KITCHEN — EQUIPMENT SCHEDULE								
EQUIP. NO.	DESIGNATION	H.P. OR K.W.	VOLTS	CONNECTION			FURNISHED BY	REMARKS
				WIRE	CONDUIT	PROT.		
1	FREEZER	1 H.P.	240 V.	No.12	3/4"	20 A	EQUIP BY OTHERS OUTLET BY CONT	24" A.F.F. ; 1φ
2	MIXER	1/3 H.P.	120 V	No.12	3/4"	20 A		48" A.F.F. ; 1φ
3	ICE MACHINE	1 H.P.	120 V	No.12	3/4"	20 A		72" A.F.F. ; 1φ
4	REFRIGERATED DISPLAY CASE	1/4 H.P.	120 V	No.12	3/4"	20 A		24" A.F.F. ; 1φ
5	RECEPTACAL		120 V	No.12	3/4	20 A		72" A.F.F. ; 1φ
6	REFRIGERATOR		120 V	No.12	3/4"	20 A		6" A.F.F. ; 1φ
7	ICE STORAGE CHEST		120 V	No.12	3/4"	20 A		39" A.F.F. ; 1φ
8	COFFEE MAKER		240 V	No.12	3/4"	20 A		39" A.F.F. ; 3φ
9	MILK DISPENSERS		120 V	No.12	3/4"	20 A		37" A.F.F. ; 1φ
10	WAFFLE MAKER		240 V	No.10	3/4	30 A		39" A.F.F. ; 3φ
11	WAFFLE MAKER		240 V	No.10	3/4"	30 A		39" A.F.F. ; 3φ
12	REFRIGERATOR	1/4 H.P.	120 V	No.12	3/4"	20 A		24" A.F.F. ; 1φ
13	TOASTER		240 V	No.12	3/4	20 A		39" A.F.F. ; 3φ
14	RECEPT.		120 V	No.12	3/4"	20 A		72" A.F.F. ; 1φ
15	MIXER		120 V	No.12	3/4"	20 A		6" A.F.F. ; 1φ
16	UPDRAFT UNIT	33.8	240 V	No.1	1 1/2"	125 A		26" A.F.F. ; 3φ
17	UP RIGHT REFRIGERATOR	1.5	120 V	No.12	3/4"	20 A		74" A.F.F. ; 1φ
18	ICE STORAGE CHEST		120 V	No.12	3/4"	20 A		39" A.F.F. ; 1φ
19	COFFEE MAKER		240 V	No.12	3/4"	20 A		39" A.F.F. ; 3φ
20	REFRIGERATOR		120 V	No.12	3/4"	20 A		6" A.F.F. ; 1φ
21	DUPLEX RECEPTACAL		120 V	No.12	3/4"	20 A		72" A.F.F. ; 1φ
22	REFRIGERATED DISPLAY CASE	1/4 H.P.	120 V	No.12	3/4"	20 A		24" A.F.F. ; 1φ
23	FLOOR RECEPTACAL		120 V	No.12	3/4"	20 A		FLOOR OUTLET ; 1φ
24	FLOOR RECEPT.		120 V	No.12	3/4"	20 A		FLOOR OUTLET ; 1φ
25	REFRIG.		120 V	No.12	3/4"	20 A		72" A.F.F. ; 1φ

Fig. 7-9. Kitchen-equipment schedule.

workmen. For example, equip. no. 1 in the schedule is a connection for a freezer. The columns in the schedules indicate that the compressor motor is 1 hp and rated for 240 volts. The circuit will consist of No. 12 wire run in ¾-inch conduit and provided with an overcurrent device rated for 20 ampere protection. The freezer itself will be furnished by others, and the outlet is to be placed 24 inches above the finished floor.

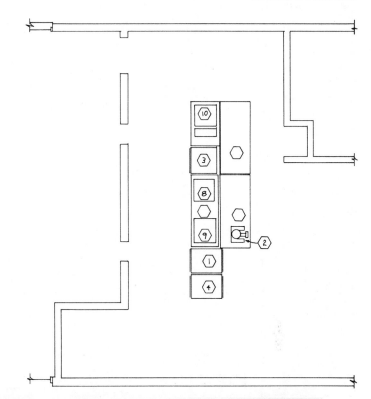

Fig. 7-10. Example showing how kitchen equipment is identified on a plan.

SYMBOL ⏻	AMP. RATING	WIRE & POLES	VOLTAGE RATING	NEMA TYPE	CONFIG- URATION	CATALOG NO *	REMARKS
A ⏻	15	3WG 2P	125	5-15R		5262	DUPLEX
B	20	3WG 2P	125	5-20R		5361 5362	SINGLE DUPLEX
C	30	3WG 2P	125	5-30R		9308	SINGLE
D	50	3WG 2P	125			9360	SINGLE
E	20	3WG 2P	250	6-20R		5461 5462	SINGLE DUPLEX
F	30	3WG 2P	250	6-30R		9330	SINGLE
G	50	3WG 2P	250	6-50R		9367	SINGLE
H	20	4WG 3P	250			7250	SINGLE 3 PHASE
J	30	4WG 3P	250			8340	SINGLE 3 PHASE
K	50	4WG 3P	250			8450	SINGLE 3 PHASE
R	50	3WG 2P	125/250	14-50R		9450	RANGE

SCHEDULE - RECEPTACLE TYPES

* HUBBEL CATALOG NOS.-FOR EXAMPLE

NOTES:

1.) ALL RECEPTACLES SHALL BE GROUNDING TYPE.

2.) SPECIAL RECEPTACLES AS CLOCK OUTLETS, WEATHERPROOF RECEPTACLES, ETC. SHALL BE AS DESCRIBED IN THE SPECIFICATIONS.

3.) VERIFY LOCATION OF ALL EQUIPMENT OUTLETS, INCLUDING HEIGHT OF RECEPTACLES & SWITCHES BEFORE ROUGHING IN.

Fig. 7-11. Schedule of receptacle types.

PANELBOARD SCHEDULE
PANEL "C"

CKT. No.	CIRCUIT BKS. POLE	FRM	TRIP	WIRE SIZE	CONNECTED LOAD IN WATTS Aφ	Bφ	Cφ	ITEMS FED OR REMARKS
1	2-P	70	30	10	2250			WATER HEATER
2	1-P		20	12	1000			LTS. & RECEPTS
3	—		—	—	—	2250		
4	1-P		15	12		1150		LIGHTS
5					1000			RECEPTS
6					1200			
7						800		RECEPTS
8						1200		
9					1200			
10					1200			
11						1200		
12						1200		
13			↓		1200			RECEPTS
14			20		1500			
15			15			1200		
16			20			1500		
17			20		1500			
18			20		1500			
19			15			1200		
20	↓	↓	20	↓		1500		↓
21								
22								
23								
24								
25								
26								
27								
28								
29								
30								
31								
32								
33								
34								
35								

Fig. 7-12. Panelboard schedule for the assignment.

SCHEDULE OF RECEPTACLE TYPES

When a variety of receptacle types are found on the working drawings, it is advantageous to organize them in a schedule, as shown in Fig. 7-11. The receptacle is used on the plans with a letter code adjacent to it. For example, if the plan shows ⊢⊙c the schedule indicates that this receptacle is rated at 30 amperes and 125 volts and that it is a 3-wire, 2-pole NEMA type 5-30R. The "Configuration" column shows the appearance of the blade slots, and the next column gives the catalog number. It can be seen from the "Remarks" column that it is a single receptacle.

SUMMARY

A schedule is a useful timesaving method of presenting notes or lists of equipment on drawings in tabular form.

With a schedule, an item can be fully described by using only a minimum of notes, whereas without a schedule, it could take pages of written specifications to properly describe the item.

A schedule, then, is an excellent means of providing essential information in a clear and accurate manner, and this allows the workmen to carry out their assignments in the least amount of time.

Only a few of the many electrical equipment schedules are given in this chapter. However, if the material given is fully understood, the reader should have no trouble in interpreting any schedule properly presented on electrical working drawings.

Other types of schedules are transformer schedule, schedule of materials, relay or remote-control schedule and motor-control schedule.

ASSIGNMENT 7

1. What is the total air-conditioning load in kVA given in the "Connected-Load Schedule" in Fig. 7-2?_____

2. What is the total connected heating load (in kilowatts) in building B in Fig. 7-3? _____

3. How many 2-pole, 20-ampere circuit breakers are called for in the panelboard schedule in Fig. 7-5? _____

4. What is the total load (in watts) connected to circuit No. 4 in Fig. 7-6? _____

5. What is the total load (in watts) connected to circuit No. 6 in the panelboard schedule of Fig. 7-12? _____

6. Which phase is the overcurrent device for circuit No. 6 connected in Fig. 7-12? _____

7. What horsepower is the motor for the mixer in the "Kitchen-Equipment Schedule" in Fig. 7-9? _____

8. What is the mounting height of receptacle number 5 in Fig? _____

9. What size conduit is indicated for the updraft unit in Fig. 7-9? _____

10. What is the voltage rating of receptacle "G" in the "Schedule—Receptacle Types" in Fig. 7-11? _____

Site Plans

A site plan is a plan view that shows the entire property with the buildings drawn in their proper location on the plot. Such plans also include sidewalks, driveways, streets, trees, and items such as water and sewer lines, electrical and telephone systems, and similar systems related to the building itself.

Site plans are drawn to scale, but in most instances, the engineer's scale is used rather than the architect's scale (which was described previously). Usually, for small buildings on small lots, a scale of $1'' = 10'$ or $1'' = 20'$ is used. This means that 1 inch (actual measurement) on the drawing is equal to 10 or 20 feet—whichever the case may be—on the ground. Since the engineer's scale is the chief means of making scaled site plans, its use should be thoroughly understood.

CIVIL ENGINEER'S SCALE

The civil engineer's scale is used fundamentally in the same manner as the architect's scale, the principal difference being that the graduations on the engineer's scale are decimal units rather than feet, as on the architect's scale.

The engineer's scale is used by placing it on the drawing with the working edge away from the user. The scale is then aligned in the direction of the required measurement. Then, by looking down over the scale, the dimension is read, in the case of an existing drawing, or the required dimension is marked off, in the case of a line that is to be drawn.

Civil engineer's scales are common in the following graduations:

$1'' = 10$ units	$1'' = 60$ units
$1'' = 20$ units	$1'' = 80$ units
$1'' = 30$ units	$1'' = 100$ units
$1'' = 40$ units	

The purpose of this scale is to transfer the relative dimensions of an object to the drawing, or vice versa.

A. _____
  $1'' = 10'$

B. _____
  $1'' = 40'$

C. _____
  $1'' = 50'$

D. _____
  $1'' = 30'$

E. _____
  $1'' = 60'$

F. _____
  $1'' = 40'$

G. _____
  $1'' = 30'$

H. _____
  $1' = 30'$

I. _____
  $1' = 20'$

J. _____
  $1'' = 40'$

K. _____
  $1'' = 60'$

L. _____
  $1'' = 10'$

Fig. 8-1. Line lengths for practice exercise in using the engineer's scale.

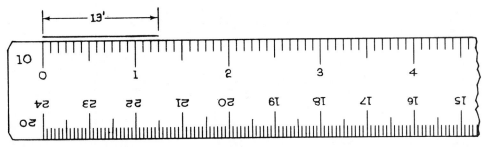

Fig. 8-2. Method of using engineer's scale to measure line dimensions.

Although the drawing itself may appear reduced in scale, depending on the size of the object and the size of the sheet to be used, the actual true-length dimensions must be shown on the drawings at all times. When you are reading or drawing plans to scale, the important point to remember is to think and speak of each dimension in its full size and not in the reduced size it happens to be on the drawing.

The practice problems in Fig. 8-1 will acquaint you with the use of the most commonly used graduations on the engineer's scale. For each problem, use the scale indicated just below the line. Determine the length of each line and write it just above the line. When all the lengths have been determined, compare your answers with the ones given at the end of this book. Fig. 8-2 shows how the first dimension is found.

DEVELOPING SITE PLANS

In general building-construction practice, it is usually the owner's responsibility to furnish the architect with property and topographic surveys, which are made by a certified land surveyor or civil engineer. These surveys will show:

1. All property lines.
2. Existing public utilities and their location on or near the property; that is, electrical lines, sanitary sewer, gas line, water-supply line, storm sewer, manholes, telephone lines, etc.

A land surveyor does the property survey (Fig. 8-3) from information obtained from a deed description of the property. A property survey shows only the property lines and their lengths as if the property were perfectly flat.

The topographic survey (Fig. 8-4) will show the property lines but, in addition, will show the physical character of the land by using contour lines, notes, and symbols. The physical characteristics may include:

1. The direction of the land slope.
2. Whether the land is flat, hilly, wooded, swampy,

high, or low, and other features of its physical nature.

All of the previous information is necessary so that the architect can properly place the building on the property. The electrical designer also needs this information to locate existing electrical utilities and to route the new service to the building.

PRACTICAL APPLICATIONS

The site plan in Fig 8-5 is that of a state-funded school. Several existing buildings are shown, but the one that is "crosshatched" is the one with which the electrical designer is concerned. This building—Swanson Hall—is to be renovated, and the electrical site plan shows the routing of the new electrical service. The new primary service is tapped from an existing transformer bank, run down an existing power pole, and then extended underground to a new pad-mounted 75-kVA transformer.

A competent electrical contractor could install this system with just the site plan, as shown in Fig. 8-5. However, the contractor would have to perform additional designs to show his workman the exact details of construction. Therefore, in order to give the contractor a better description of the installation, the electrical designer provides additional electrical details.

Since the work involved actually begins at the existing transformer bank, a drawing entitled "Pole Detail" was added to the working drawings (Fig. 8-6). This detail leaves little doubt about exactly what is required of the contractor. For example, the detail shows the existing crossarm, pole, and 12.5-kV primary service conductors, indicating that the electrical contractor will tap the existing service conductors, which are then run to three new fuse cutouts (fused at 10 amperes). The circuit continues on to new cable terminals and lightning arresters. A new three-wire shielded cable is then connected to the cable terminators and run down the pole. Even the connection of the ground wire and the sizes of machine bolts, thru bolts, lag bolts and the ground bolt are given.

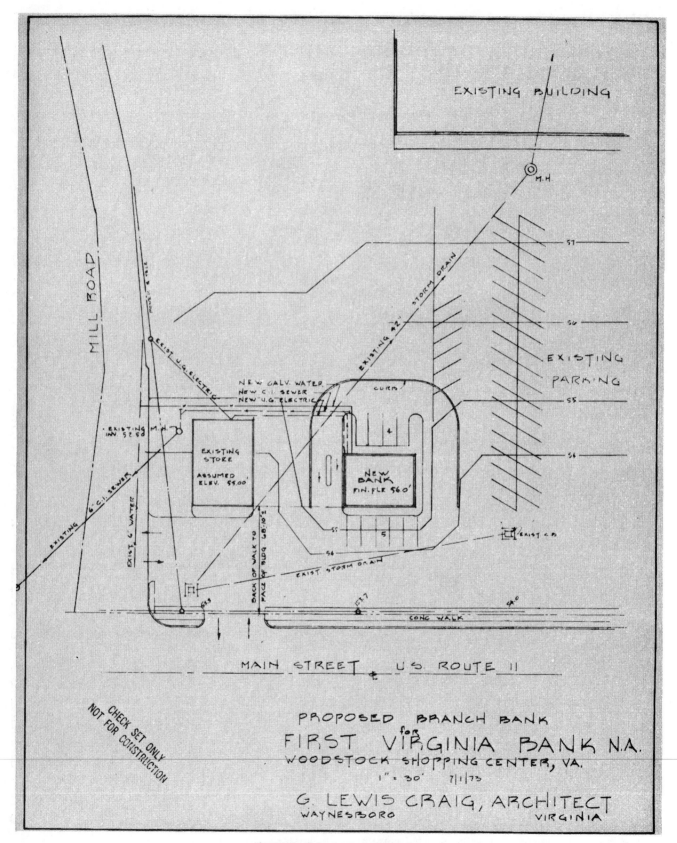

Fig. 8-3. Typical property survey plan.

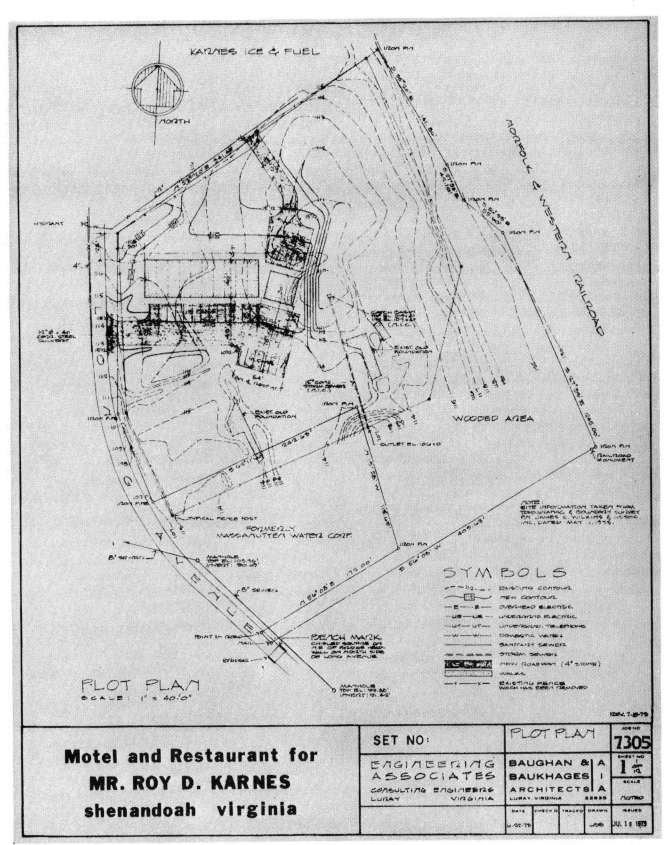

Fig. 8-4. Typical topographical plan.

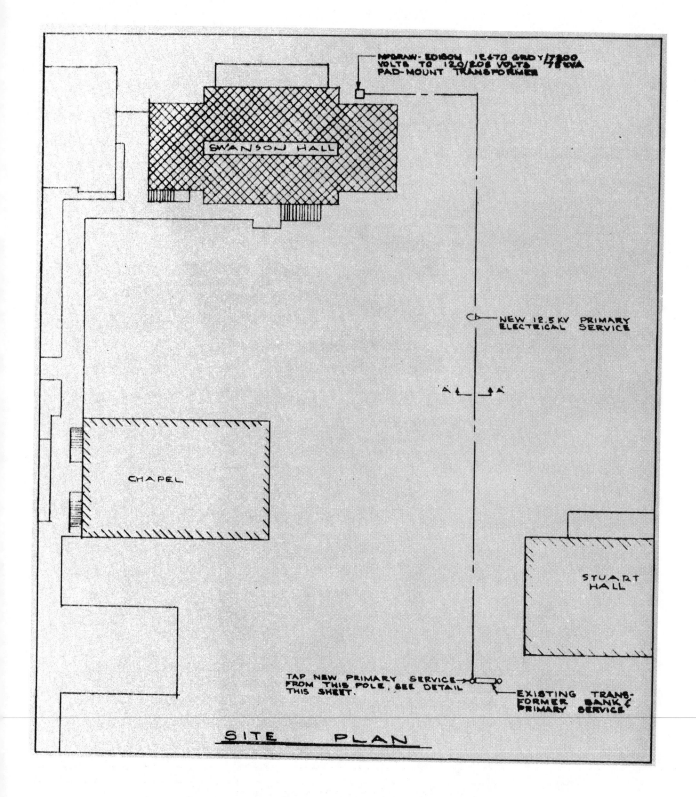

Fig. 8-5. Electrical site plan for a building project showing electrical system.

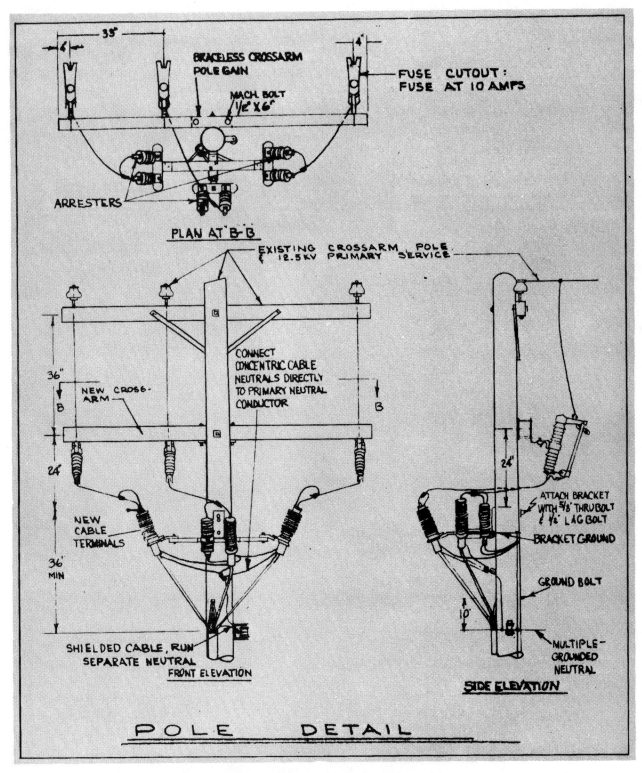

Fig. 8-6. Pole detail for use with the site plan in Fig. 8-5.

The cable then runs down the existing power pole to a given distance below the ground, extending to the transformer pad. Fig. 8-7 shows section A-A of this underground cable and again clearly shows what is expected of the contractor. The contractor can see that the minimum depth of the trench from grade level to the top of the concrete fill is 42 inches. The minimum concrete fill is 2 feet; the fiberduct is 4 inches/2 inches = 2 inches, plus 3 inches of sand fill. This means that the minimum trench depth is $42 + 2 + 2 + 3) = 49$ inches. After the trench is dug, a 3-inch layer of sand fill is placed in the bottom of the trench. Next, 3-conductor, No. 4 AWG, 15-kV cable is laid on the sand fill and covered with 4-inch fiberduct that has been split (sawed) lengthwise. Concrete is then poured over the protected conductors, and partial backfill with dirt is started. About 21 inches below grade, a continuous warning ribbon is put in the trench to warn those who may be digging there in the future that high-voltage cable is buried beneath this area.

A note lettered on the drawings or written in the specifications will give additional data pertaining to the buried cable. The note for the project in question reads as follows:

NOTE: All high-voltage underground wiring shall be directly buried, 15 kV, No. 4 AWG, aluminum type AA, 3-conductor, 7-strand phase conductors with extruded semiconducting cross-linked polyethylene strand shielding, 175 mils of cross-linked polyethylene insulation, 30 mils semiconducting polyethylene jacket and 10 No. 14 AWG base copper concentric neutral and buried at a minimum depth of 42 inches.

As mentioned previously, the underground primary service conductor continues on to a pad-mounted transformer. At this point, the designer felt that additional drawing details were needed. Fig. 8-8 shows the desired detail. This detail gives the construction of the transformer pad, the grounding, and the conduit entries to the transformer connection compartment. Referring back to Fig. 8-5, we find that a McGraw-Edison 12470 GRDY/7200 volts to 120/208 volts, 75-kVA transformer is specified.

The following should also appear either in the written specifications or as notes on the working drawings in order to complete the electrical site-work design:

UNDERGROUND DISTRIBUTION

1. To avoid physical obstructions and to provide adequate space separation for fire protection, the following minimum clearances are given for locating transformer foundations:
   a. 10 feet from window (along wall horizontally).
   b. 10 feet below window (vertically).
   c. 5 feet from building or other structure.
   d. 10 feet from door or entrance (along wall).
   e. 10 feet from fire escape.
   f. 10 feet from ventilating ducts.
2. Concrete pad may be poured in place or may be precast.
3. All conduits shall be installed before placing pad. Conduits should not be placed under the section of the pad supporting the transformer so that the original ground is not disturbed.
4. Conduit shall be rigid polyvinyl chloride.

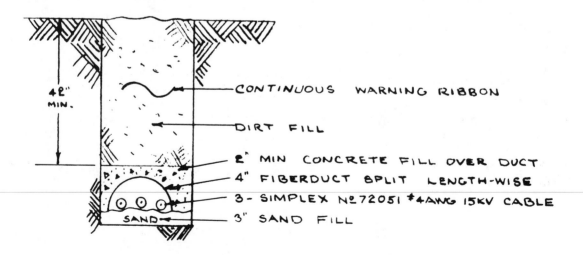

Fig. 8-7. Section A-A of the underground cable in Fig. 8-5.

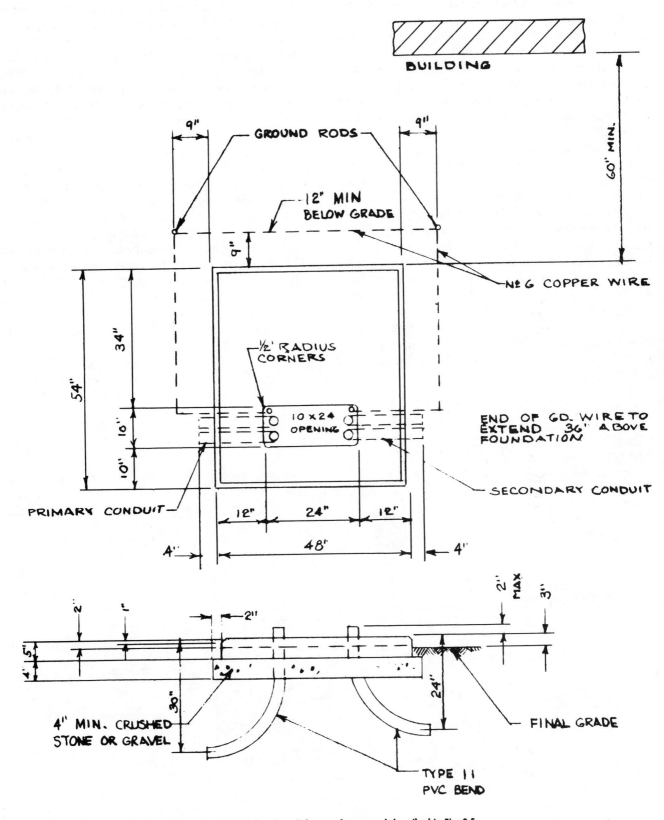

Fig. 8-8. Detail drawing of the transformer pad described in Fig. 8-5.

5. The crushed stone or gravel shall be thoroughly compacted.
6. To prevent water migration from the concrete, a waterproof membrane shall be placed on the crushed stone or gravel before the concrete is poured.
7. Backfill shall be clean granular soil, free of large stones and perishable material. All backfill shall be spread and compacted in maximum layers of 8 inches.
8. Where damage to the transformer by vehicles is possible, the transformer shall be protected by an appropriate barrier.
9. All spare conduits and openings shall be sealed to prevent the entry of rodents and other animals into the transformer compartment.
10. If conduit extends into the building, it must be sealed at the building end to prevent gas from entering the building through the conduit. Use AQUASEAL stock number 594006 for the sealing compound.
11. The concrete shall develop 3000 psi at 28 days age, contain a minimum of 5.5 bags of cement per cubic yard and a maximum of 6 gallons of water per 94-pound bag of cement, and conform to ASTM designation C-94.
12. Reinforcing steel shall be used and shall conform to ASTM designations.

The power-riser diagram in Fig. 8-9 shows the secondary (low-voltage) service extending from the transformer to panelboard LB inside the building. Notice that four 500 MCM THW copper (Cu) conductors have been specified to be pulled in 3½-inch conduit. The remaining panels in the building are fed from panel LB.

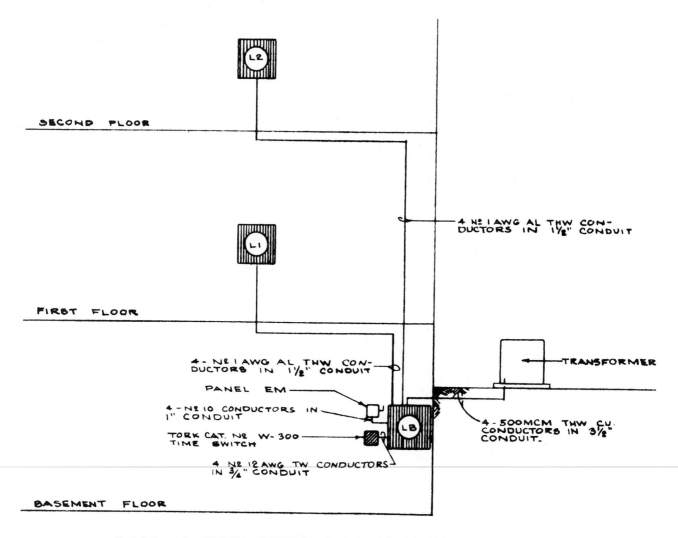

Fig. 8-9. Power-riser diagram showing secondary electrical connections for the building in Fig. 8-5.

SUMMARY

A site plan is a plan view that shows the entire property with the buildings drawn in their proper location on the plot.

Site plans are drawn to scale, and the engineer's scale is most often used for this purpose.

When reading plans that are drawn to scale, remember to think and speak of each dimension in its full size rather than in the reduced size that it appears on the drawing.

ASSIGNMENT 8

The site plan in Fig. 8-10 is typical of the plans encountered by those who must interpret construction drawings. Examine this plan carefully; then answer the following questions and fill in the blanks—using Fig. 8-10 as a reference.

1. The symbol ▸•◂ represents lighting standards for parking lights. How many of these lighting standards are shown on the drawing? _____

2. The symbol ⊙ represents a manhole for electrical connections; how many of these are shown n the drawing? _____

3. The rectangular areas resembling brick patterns represent _____ areas on the drawings.

4. The symbol ⊓ means square feet. Give the area, in square feet, for the following:
    A. No. 801E _____
    B. No. 801D _____
    C. No. 801C _____
    D. No. 801B _____
    E. No. 801A _____

5. According to the drawing, by how many feet will the existing entrance be widened? _____

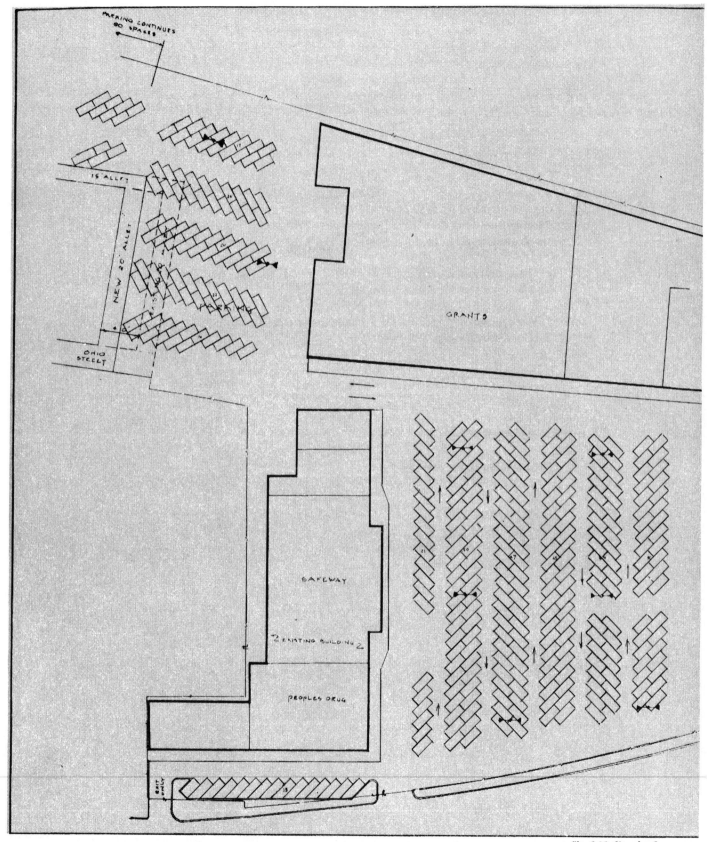

Fig. 8-10. Site plan for

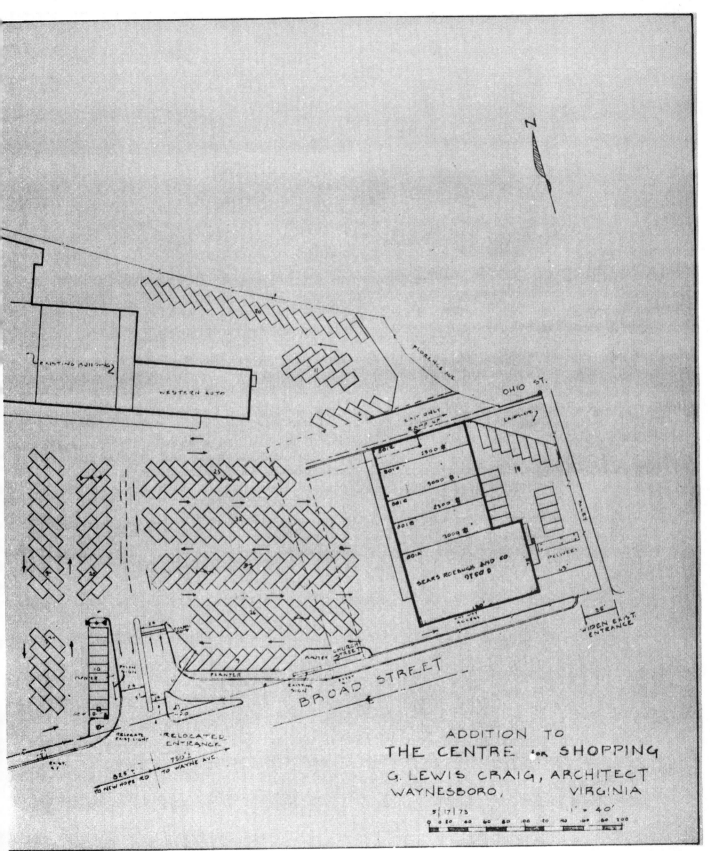

ADDITION TO
THE CENTRE for SHOPPING
G. LEWIS CRAIG, ARCHITECT
WAYNESBORO,   VIRGINIA

5/17/73   1" = 40'

use with assignment.

9

Electrical Specifications

The specifications for a building or project are the written description of what will be required by the owner, architect, and engineer. Together with the working drawings, the specifications form the basis of the contract requirements for the construction.

Those who must interpret construction drawings and specifications must always be on the alert for conflicts existing between different sheets of contract drawings or between the working drawings and the written specifications. Such conflicts occur particularly when:

1. Standard or prototype specifications are used in conjunction with specific working drawings.
2. Standard or previously prepared drawings are changed or amended by reference in the specifications only and, for some reason of the architect, owner, or engineer, the drawings themselves are not changed.

In such instances, it is the responsibility of the person in charge of the project to ascertain which one takes precedence over the other—the drawings or the specifications. When such a situation exists, the matter must be cleared up, preferably before the work is installed, in order to avoid added cost to either the owner or the contractor.

In general, electrical specifications give the grade of materials to be used on the project and the manner in which the electrical system shall be installed. The following sample illustrates the general wording and contents of a typical electrical specification actually used on a commercial project. Although its contents need not be memorized, the entire specification should be read through to give the reader an idea of what the contents are. The exercises at the end of this chapter are designed to help the reader learn to use written specifications with ease.

DIVISION 16 - ELECTRICAL

SECTION 16A - GENERAL PROVISIONS

1. Portions of the sections of the Documents designated by the letters "A", "B" & "C" and "DIVISION ONE - GENERAL REQUIREMENTS" apply to this Division.

2. Consult Index to be certain that set of Documents and Specifications is complete. Report omissions or discrepancies to the Architect.

3. SCOPE OF THE WORK:

a. The scope of the work consists of the furnishing and installing of complete electrical systems - exterior and interior - including miscellaneous systems. The Electrical Contractor shall provide all supervision, labor, materials, equipment, machinery, and any and all other items necessary to complete the systems. The Electrical Contractor shall note that all items of equipment are specified in the singular; however, the Contractor shall provide and install the number of items of equipment as indicated on the drawings and as required for complete systems.

b. It is the intention of the Specifications and Drawings to call for finished work, tested, and ready for operation.

c. Any apparatus, appliance, material or work not shown on drawings but mentioned in the specifications, or vice versa, or any incidental accessories necessary to make the work complete and perfect in all respects and ready for operation, even if not particularly specified, shall be furnished, delivered and installed by the Contractor without additional expense to the Owner.

d. Minor details not usually shown or specified, but necessary for proper installation and operation, shall be included in the Contractor's estimate, the same as if herein specified or shown.

e. With submission of bid, the Electrical Contractor shall given written notice to the Architect of any materials or apparatus believed inadequate or unsuitable, in violation of laws, ordinances, rules; and any necessary items or work omitted. In the absence of such written notice, it is mutually agreed the Contractor has included the cost of all required items in his proposal, and that he will be responsible for the approved satisfactory functioning of the entire system without extra compensation.

4. ELECTRICAL DRAWINGS:

a. The Electrical drawings are diagrammatic and indicate the general arrangement of fixtures, equipment and work included in the contract. Consult the Architectural drawings and details for exact location of fixtures and equipment; where same are not definitely located, obtain this information from the Architect.

b. Contractor shall follow drawings in laying out work and check drawings of other trades to verify spaces in which work will be installed. Maintain maximum headroom and space conditions at all points. Where headroom or space conditions appear inadequate, the Architect shall be notified before proceeding with installation.

c. If directed by the Architect, the Contractor shall, without extra charge, make reasonable modifications in the layout as needed to prevent conflict with work of other trades or for proper execution of the work.

5. CODES, PERMITS AND FEES:

a. Contractor shall give all necessary notices, including electric and telephone utilities, obtain all permits and pay all government taxes, fees and other costs, including utility connections or extensions, in connection with his work; file all necessary plans, prepare all documents and obtain all necessary approvals of all governmental departments having jurisdiction; obtain all required certificates of inspection for his work and deliver same to the Architect before request for acceptance and final payment for the work.

b. Contractor shall include in the work, without extra
cost to the Owner, any labor, materials, services, apparatus, drawings (in addition to contract drawings and documents) in order to comply with all applicable laws,
ordinances, rules and regulations, whether or not shown
on drawings and/or specified.

c. Work and materials shall conform to the latest rules
of the National Board of Fire Underwriters' Code, Regulations of the State Fire Marshal, and with applicable
local codes and with all prevailing rules and regulations
pertaining to adequate protection and/or guarding of any
moving parts, or otherwise hazardous conditions. Nothing
in these specifications shall be construed to permit work
not conforming to the most stringent of applicable codes.

d. The National Electric Code, the Local Electric Code,
and the electrical requirements as established by the State
and Local Fire Marshal, and rules and regulations of the
power company serving the project, are hereby made part of
this specification. Should any changes be necessary in
the drawings or specifications to make the work comply
with these requirements, the Electrical Contractor shall
notify the Architect.

6. SHOP DRAWINGS:

a. The Electrical Contractor shall submit five (5) copies
of the shop drawings to the Architect for approval within
thirty (30) days after the award of the general contract.
If such a schedule cannot be met, the Electrical Contractor
may request in writing for an extension of time to the
Architect. If the Electrical Contractor does not submit
shop drawings in the prescribed time, the Architect has the
right to select the equipment.

b. Shop drawings shall be submitted on all major pieces
of electrical equipment, including service entrance equipment, lighting fixtures, panel boards, switches, wiring
devices and plates and equipment for miscellaneous systems.
Each item of equipment proposed, shall be a standard catalog
product of an established manufacturer. The shop drawing
shall give complete information on the proposed equipment.
Each item of the shop drawings shall be properly labeled,
indicating the intended service of the material, the job
name and Electrical Contractor's name.

c. The shop drawings shall be neatly bound in five (5) sets and submitted to the Architect with a letter of transmittal. The letter of transmittal shall list each item submitted along with the manufacturer's name.

d. Approval rendered on shop drawings shall not be considered as a guarantee of measurements or building conditions. Where drawings are approved, said approval does not mean that drawings have been checked in detail; said approval does not in any way relieve the Contractor from his responsibility, or necessity of furnishing material or performing work as required by the contract drawings and specifications.

7. <u>COOPERATION WITH OTHER TRADES</u>:

a. The Electrical Contractor shall give full cooperation to other trades and shall furnish (in writing, with copies to Architect) any information necessary to permit the work of all trades to be installed satisfactorily and with least possible interference or delay.

b. Where the work of the Electrical Contractor will be installed in close proximity to work of other trades, or where there is evidence that the work of the Electrical Contractor will interfere with the work of other trades, he shall assist in working out space conditions to make a satisfactory adjustment. If so directed by the Architect, the Electrical Contractor shall prepare composite working drawings and sections at a suitable scale clearly showing how his work is to be installed in relation to the work of other trades. If the Electrical Contractor installs his work before coordinating with other trades or so as to cause any interference with work of other trades, he shall make necessary changes in his work to correct the condition without extra charge.

c. The complexity of equipment and the variation between equipment manufacturers requires complete coordination of all trades. The Contractor, who offers for consideration, substitutes of equal products of reliable manufacturers, has to be responsible for all changes that effect his installation and the installation and equipment of other trades. All systems and their associated controls must be completely installed, connected, and operating to the satisfaction of the Architect prior to final acceptance and contract payment.

8. TEMPORARY ELECTRICAL SERVICE:

a. The Electrical Contractor shall be responsible for all arrangements and costs for providing at the site, temporary electrical metering, main switches and distribution panels as required for construction purposes. The distribution panels shall be located at a central point designated by the Architect. The General Contractor shall indicate prior to installation whether three phase or single phase service is required.

9. ELECTRICAL CONNECTIONS:

a. The Electrical Contractor shall provide and install power wiring to all motors and electrical equipment complete and ready for operation including disconnect switches and fuses. Starters, relays and accessories shall be furnished by others unless otherwise noted, but shall be installed by the Electrical Contractor. This Contractor shall be responsible for checking the shop drawings of the equipment manufacturer to obtain the exact location of the electrical rough-in and connections for equipment installed.

b. The Mechanical Contractor will furnish and install all temperature control wiring and all interlock wiring unless otherwise noted.

c. It shall be the responsibility of the Electrical Contractor to check all motors for proper rotation.

10. AS-BUILT DRAWINGS:

a. The Electrical Contractor shall maintain accurate records of all deviations in work as actually installed from work indicated on the drawings. On completion of the project, two (2) complete sets of marked-up prints shall be delivered to the Architect.

11. INSPECTION AND CERTIFICATES:

a. On the completion of the entire installation, the approval of the Architect and Owner shall be secured, covering the installation throughout. The Contractor shall obtain and pay for Certificate of Approval from the public authorities having jurisdiction. A final inspection certificate shall be submitted to the Architect prior to final payment. Any and all cost incurred for fees shall be paid for by the Contractor.

12. TESTS:

a. The right is reserved to inspect and test any portion of the equipment and/or materials during the progress of its erection. This Contractor shall test all wiring and connections for continuity and grounds, before connecting any fixtures or equipment.

b. The Contractor shall test the entire system in the presence of the Architect or his engineer when the work is finally completed to insure that all portions are free from short circuits or grounds. All equipment necessary to conduct these tests shall be furnished at the Contractor's expense.

13. EQUIVALENTS:

a. When material or equipment is mentioned by name, it shall form the basis of the Contract. When approved by the Architect in writing, other material and equipment may be used in place of those specified, but written application for such substitutions shall be made to the Architect as described in the Bidding Documents. The difference in cost of substitute material or equipment shall be given when making such request. Approval of substitute is, of course, contingent on same meeting specified requirements and being of such design and dimensions as to comply with space requirements.

14. GUARANTEE:

a. The Electrical Contractor shall guarantee, by his acceptance of the contract, that all work installed will be free from defects in workmanship and materials. If during the period of one year, or as otherwise specified, from date of Certificate of Completion and acceptance of work, any such defects in workmanship, materials or performance appear, the Contractor shall, without cost to the Owner, remedy such defects within a reasonable time to be specified in notice from Architect. In default, the Owner may have such work done and charge cost to Contractor.

SECTION 16B - BASIC MATERIALS AND WORKMANSHIP

1. Portions of the sections of the Documents designated by the letters "A", "B" & "C" and "DIVISION ONE - GENERAL REQUIREMENTS" apply to this Division.

2. Consult Index to be certain that set of Documents and Specifications is complete. Report omissions or discrepancies to the Architect.

3. CONDUIT MATERIAL AND WORKMANSHIP:

    a. GENERAL: The Electrical Contractor shall install a complete raceway system as shown on the drawings and stated in other sections of the specifications. All material used in the raceway system shall be new and the proper material for the job. Conduit, couplings and connectors shall be a product of a reputable manufacturer equal to conduit as manufactured by Triangle Conduit and Cable or National Electric.

    b. CONDUIT INSTALLATION:

        (1) Conduit shall be of ample size to permit the ready insertion and withdrawal of conductors without abrasion. All joints shall be cut square, reamed smooth and drawn up tight.

        (2) Concealed conduits shall be on run in as direct a manner and with as long a bend as possible. Exposed conduit shall be run parallel to or at right angles with the lines of the building. All bends shall be made with standard ells, conduit bent to a radius not less than shown in N.E.C., or screw jointed conduit fittings. All bends shall be free of dents or flattening. Not more than the equivalent of four quarter bends shall be used in any run between terminals and cabinets, or between outlets and junction or pull boxes.

        (3) Conduits shall be continuous from outlet to outlet and from outlet to cabinets, junction or pull boxes, and shall enter and be secured at all boxes in such a manner that each system shall be electrically continuous throughout.

        (4) A #14 galvanized iron or steel fish wire shall be left in all conduits in which the permanent wiring is not installed.

(5)   Where conduits cross building joints, furnish and install O.Z. Electric Manufacturing Company expansion fittings for contraction, expansion and settlement.

(6)   Open ends shall be capped with approved manufactured conduit seals as soon as installed and kept capped until ready to install conductors.

(7)   Conduit shall be securely fastened to all sheet metal outlets, junction and pull boxes with galvanized lock-nuts and bushings, care being observed to see that the full number threads project through to permit the bushings to be drawn tight against the end of conduit, after which the lock-nut shall be made up sufficiently tight to draw the bushings into firm electrical contact with the box.

(8)   For all flush-mounted panels there shall be provided and installed 1¼" empty conduit up through wall and turned out above ceiling and one 1¼" empty conduit down to space below floor except where slab is on grade.

c.   CONDUIT HANGERS AND SUPPORTS:

(1)   Conduit throughout the project shall be securely and rigidly supported to the building structure in a neat and workmanlike manner and wherever possible, parallel runs of horizontal conduit shall be grouped together on adjustable trapeze hangers. Support spacing shall not be more than eight feet.

(2)   Exposed conduit shall be supported by one hole malleable iron straps, two hole straps, suitable beam clamps or split ring conduit hanger with support rod.

(3)   Single conduit 1¼" and larger run concealed horizontally shall be supported by suitable beam clamps or split ring conduit hangers with support rod. Multiple runs of conduit shall be grouped together on trapeze hangers where possible. Vertical runs shall be supported by steel riser clamps.

(4)   Conduit one inch and smaller run concealed above a ceiling may be supported directly to the building structure with strap hangers or No. 14 ga. galvanized wire provided the support spacing does not exceed four feet.

4. OUTLET BOXES:

a. GENERAL:

(1) Before locating the outlet boxes check all of the architectural drawings for type of construction and to make sure that there is no conflict with other equipment. The outlet boxes shall be symmetrically located according to room layout and shall not interfere with other work or equipment. Also note any detail of the outlets shown on the drawings.

(2) Outlet boxes shall be made of galvanized sheet steel unless otherwise noted or required by the N.E.C. and shall be of the proper code size for the required number of conductors. Outlet boxes shall be a minimum of 4 inches square unless specifically noted on the drawings with the exception of a box containing only two current carrying conductors may be smaller. The outlet boxes shall be complete with the approved type of connectors and required accessories.

(3) The outlet boxes shall be complete with raised device covers as required to accept device installed. All outlet boxes must be securely fastened in position with the exposed edge of the raised device cover set flush with the finished surface. Approved factory made knockouts seals shall be installed where knockouts are not intact. Galvanized outlet boxes shall be manufactured by RACO, STEEL CITY, APPLETON or approved equal.

(4) Outlet boxes for exposed work shall be handy boxes with handy box covers unless otherwise noted.

(5) Outlet boxes located on the exterior in damp or wet locations or as otherwise noted shall be threaded cast aluminum device boxes such as CROUSE HINDS Type "FS" or "FD".

b. RECEPTACLE OUTLET BOXES: Wall receptacles shall be mounted approximately 18" above the finished floor (AFF) unless otherwise noted. When the receptacle is mounted in a masonry wall the bottom of the outlet box shall be in line with the bottom of a masonry unit. Receptacles for electric water coolers shall be installed behind the coolers in accordance with manufacturers recommendations. All receptacle outlet boxes shall be equipped with grounding lead which shall be connected to grounding terminal of the device.

c. SWITCH OUTLET BOXES: Wall switches shall be mounted approximately 54 inches above the finished floor (AFF) unless otherwise noted. When the switch is mounted in a masonry wall the bottom of the outlet box shall be in line with the bottom of a masonry unit. Where more than two switches are located, the switches shall be mounted in a gang outlet box with gang cover. Dimmer switches shall be individually mounted unless otherwise noted. Switches with pilot lights, switches with overload motor protection and other special switches that will not conveniently fit under gang wall plates may be individually mounted.

d. LIGHTING FIXTURE OUTLET BOXES: The lighting fixture outlet boxes shall be furnished with the necessary accessories to install the fixture. The supports must be such as not to depend on the outlet box supporting the fixture. The supports for the lighting fixture shall be independent of the ceiling system. All ceiling outlet boxes shall be equipped with raised circular cover plates with its edge set flush with surface of finished ceiling.

5. PULL BOXES:

a. Pull boxes shall be installed at all necessary points, whether indicated on the drawings or not to prevent injury to the insulation or other damage that might result from pulling resistance for other reasons necessary to proper installation. Pull box locations shall be approved by the Architect prior to installation. Minimum dimensions shall be not less than N.E.C. requirements and shall be increased if necessary for practical reasons or where required to fit a job condition.

b. All pull boxes shall be constructed of galvanized sheet steel, code gauge, except that no less than 12 gauge shall be used for any box.

c. Where boxes are used in connection with exposed conduit, plain covers attached to the box with a suitable number of counter-sunk flat head machine screws may be used.

d. Where so indicated, certain pull boxes shall be provided with barriers. These pull boxes shall have a single cover plate, and the barriers shall be of the same gauge as the pull box.

e. Each circuit in pull box shall be marked with a tag guide denoting panels to which they connect.

f. Exposed pull boxes will not be permitted in the public spaces.

6. <u>WIREWAYS OR WIRE TROUGHS</u>:

a. Wireways shall be used where indicated on the drawings and for mounting groups of switches and/or starters. Wireways shall be the standard manufactured product of a company regularly producing wireway and shall not be a local shop assembled unit. Wireway shall be of the hinged cover type, Underwriters' listed, and of sizes indicated or as required by N.E.C. Finish shall be medium light gray enamel over rush inhibitor. Wireways shall be of the rain-tight construction where required. Wireways shall be General Electric Type HS or approved equal.

7. <u>CONDUCTOR MATERIAL AND WORKMANSHIP</u>:

a. GENERAL:

(1) The Electric Contractor shall provide and install a complete wiring system as shown on the drawing or specifications herein. All conductors used in the wiring system, shall be soft drawn copper wire having a conductivity of not less than 98% of that of pure copper, with 600-volt rating, unless otherwise noted. Wire shall be as manufactured by General Cable, Triangle or approved equal.

(2) The wire shall be delivered to the site in their original unbroken packages, plainly marked or tagged as follows: (a) Underwriters' Labels (b) Size, kind and insulation of the wire (c) Name of manufacturing company and the trade name of the wire.

b. CONDUCTOR WORKMANSHIP:

(1) Install conductors in all raceways as required, unless otherwise noted, in a neat and workmanlike manner. Telephone conduits and empty conduits as noted, shall have a No. 14 ga. galvanized pull wire left in place for future use.

(2)   Conductors shall be color coded in accordance with the National Electric Code. Mains, feeders, sub-feeders shall be tagged in all pull, junction and outlet boxes and in the gutter of panels with approved code type wire markers.

(3)   No lubricant other than powdered soapstone or approved pulling compound may be used to pull conductors.

(4)   At least eight (8) inches of slack wire shall be left in every outlet box whether it be in use or left for future use.

(5)   All conductors and connections shall test free of grounds, shorts and opens before turning the job over to the Owner.

8.   LUGS, TAPS AND SPLICES:

a.   Joints on branch circuits shall occur only where such circuits divide and shall consist of one through circuit to which shall be spliced the branch from the circuit. In no case shall joints in branch circuits be left for the fixture hanger to make. No splices shall be made in conductor except at outlet boxes, junction boxes, or splice boxes.

(1)   All joints or splices for No. 10 AWG or smaller shall be made with UL approved wire nuts or compression type connectors.

(2)   All joints or splices for No. 8 AWG or larger shall be made with a mechanical compression connector. After the conductors have been made mechanically and electrically secure, the entire joint or splice shall be covered with Scotch No. 33 tape or approved equal to make the insulation of the joint or splice equal to the insulation of the conductors. The connector shall be UL approved.

9.   ACCESS DOORS:

a.   The Electrical Contractor shall furnish to the lather the access doors as shown on the drawings or required for access to junction boxes, etc. The doors shall be 12" square, unless otherwise noted, hinged metal door with metal frames.

Door and frame shall be not lighter than 16 gauge sheet steel. The access door shall be of the flush type with screwdriver latching device. The frame shall be constructed so that it can be secured to building material as required. The access doors shall be Milcor or equal. Access door and location shall meet the approval of the Architect.

10. FUSES:

a. Fuses manufactured by Buss or Shawmut shall be furnished and installed as required. Motor protection fuses shall be dual element.

11. CUTTING AND PATCHING:

a. On new work the Electrical Contractor shall furnish sketches to the General Contractor showing the locations and sizes of all openings, chases, and furnish and locate all sleeves and inserts required for the installation of the electrical work before the walls, floors and roof are built. The Electrical Contractor shall be responsible for the cost of cutting and patching where any electrical items were not installed or where incorrectly sized or located. The Contractor shall do all drilling required for the installation of his hangers.

b. On alterations and additions to existing projects, the Electrical Contractor shall be responsible for the cost of all cutting and patching, unless otherwise noted.

c. No structural members shall be cut without the approval of the Architect, and all such cutting shall be done in a manner directed by him. All patching shall be performed in a neat and workmanlike manner acceptable to the Architect.

12. EXCAVATION AND BACKFILLING:

a. The Electrical Contractor shall be responsible for excavation, backfill, tamping, shoring, bracing, pumping, street cuts, repairing of finished surface and all protection for safety of persons and property as required for installing a complete electrical system. All excavation and backfill shall conform to the Architectural Section of the specifications.

b.   Excavation shall be made in a manner to provide a uniform bearing for conduit. Where rock is encountered, excavate 3" below conduit grade and fill with gravel to grade.

c.   After required test and inspections, backfill the ditch and tamp. The first foot above the conduit shall be hand backfilled with rock-free clean earth. The backfill in the ditches on the exterior and interior of the building shall be tamped to 90%. The Electrical Contractor shall be responsible for any ditches that go down.

13.   EQUIPMENT AND INSTALLATION WORKMANSHIP:

a.   All equipment and material shall be new and shall bear the manufacturer's name and trade name. The equipment and material shall be essentially the standard product of a manufacturer regularly engaged in the production of the required type of equipment and shall be the manufacturer's latest approved design.

b.   The Electrical Contractor shall receive and properly store the equipment and material pertaining to the electrical work. The equipment shall be tightly covered and protected against dirt, water, chemical or mechanical injury and theft. The manufacturer's directions shall be followed completely in the delivery, storage, protection and installation of all equipment and materials.

c.   The Electrical Contractor shall provide and install all items necessary for the complete installation of the equipment as recommended or as required by the manufacturer of the equipment or required by code without additional cost to the Owner, regardless whether the items are shown on the plans or covered in the Specifications.

d.   It shall be the responsibility of the Electrical Contractor to clean the electrical equipment, make necessary adjustments and place the equipment into operation before turning equipment over to Owner. Any paint that was scratched during construction shall be "touched-up" with factory color paint to the satisfaction of the Architect. Any items that were damaged during construction shall be replaced.

14. CONCRETE PADS, SUPPORTS AND ENCASEMENT:

a.   The Electrical Contractor shall be responsible for
all concrete pads, supports, piers, bases, foundations
and encasement required for the electrical equipment and
conduit.   The concrete pads for the electrical equipment
shall be six (6) inches larger all around than the base of
the equipment and a minimum of 4 inches thick unless
specifically indicated otherwise.

15. WATERPROOFING:

a.   The Electrical Contractor shall provide all flashing,
caulking and sleeves required where his items pass thru
the outside walls or roof.   The Waterproofing of the
openings shall be made absolutely watertight.   The method
of installation shall conform to the requirements of
Division 7 - Moisture Control and/or meet the approval
of the Architect.

SECTION 16C - SERVICE ENTRANCE SYSTEM

1. Portions of the sections of the Documents designated by the letters "A", "B" & "C" and "DIVISION ONE - GENERAL REQUIREMENTS" apply to this Division.

2. Consult Index to be certain that set of Documents and Specifications is complete. Report omissions or discrepancies to the Architect.

3. SERVICE ENTRANCE:

    a. GENERAL: The Electrical Contractor shall provide and install a complete service entrance system as shown on the drawings or as required for a complete system. All material and workmanship shall conform with Section 16B of the specifications, National Electric Code and the electric code. The electric service entrance shall conform to the requirements and regulations of the electric utility serving the project.

    b. ELECTRIC UTILITY CHARGE: The Electrical Contractor shall make all arrangements with the electric utility and pay all charges made by the electric utility for permanent electric service to the project. In the event that the electric utility's charges are not available at the time the project is bid, the Electrical Contractor shall qualify his bid to notify the Owner that such charges are not included.

    c. METERING: The Electrical Contractor shall provide and install raceway, install current transformer cabinet and/or meter trim, for metering facilities as required by the electric utility serving the project. The electric utility will provide the meter installation including meter, current transformers and connections.

    d. GROUNDING: The Electrical Contractor shall properly ground the electrical system as required by the National Electrical Code.

    e. CONDUIT: The conduit used for service entrance shall be galvanized rigid steel conduit unless otherwise noted on drawings.

f.   CONDUCTORS: Conductors for the service entrance shall be copper type RHW or THW rated at 75°C unless otherwise noted. The conductors indicated on the drawings are based on aluminum.

SECTION 16D - ELECTRICAL DISTRIBUTION SYSTEM

1. Portions of the sections of the Documents designated by the letters "A", "B" & "C" and "DIVISION ONE - GENERAL REQUIREMENTS" apply to this Division.

2. Consult Index to be certain that set of Documents and Specifications is complete. Report omissions or discrepancies to the Architect.

3. FEEDERS AND BRANCH CIRCUITS:

a. GENERAL: The Electrical Contractor shall provide and install a complete electrical distribution system as shown on the drawings or as required for a complete system. All materials and workmanship shall conform with Section 16B of the Specifications, National Electric Code and the local electric code.

b. CONDUIT MATERIALS:

(1) Rigid Conduit (Heavy Wall): Rigid conduit shall be galvanized rigid steel conduit with a minimum size of 3/4 inch unless otherwise noted. Rigid steel conduit shall be installed for the following services and locations: service entrance, underground in contact with earth, in concrete slab, panel feeders, exterior of building walls, motor feeders over 10 HP, electrical equipment feeders over 10 KW, "wet" locations, and as required by the National Electric Code and local codes.

(2) Electrical Metallic Tubing (EMT): Electrical metallic tubing shall be galvanized steel with a minimum size of 3/4 inch. Electrical metallic tubing shall be used in all locations not otherwise specified for rigid or flexible conduit and where not in violation of the National Electric Code.

(3) Flexible Metal Conduit: Flexible metal conduit shall be galvanized steel. Flexible metal conduit located in wet locations, shall be the Liquid-Tight type. Flexible metal conduit may be used in place of EMT where completely accessible, such as above removable acoustical tile ceilings and for exposed work in unfinished spaces.

A short piece of flexible metal conduit shall be used for the connection to all motors and vibrating equipment, connection between recessed light fixtures and junction box, and as otherwise noted, provided the use meets the requirements of the National Electric Code and local codes. The flexible metal conduit shall be the type approved for continuous grounding.

c. CONDUCTOR MATERIAL:

(1) The conductor material shall be as follows, unless otherwise noted:

(a) <u>Feeders</u>: Shall be Type RHW or THW rated at 75°C.

(b) <u>Branch Circuits</u>: Shall be Type THW rated at 75°C, except branch circuits with conductor sizes of No. 10 and smaller in dry locations may be Type TW rated at 60°C.

(c) <u>Special Locations</u>: Conductors in special locations such as range hoods, lighting fixtures, etc., shall be as required by the National Electrical Code, local code or as otherwise noted.

(2) No Conductor shall be smaller than No. 12 wire, except for the control wiring and as stated in other sections of the Specifications or on the drawings. Wiring to switches shall not be considered as control wiring.

(3) Conductors indicated on the drawings are based on copper. Panel, motor ard electrical equipment feeders with a size of No. 1/0 and larger may be aluminum, providing the size of the conductor is increased to have the same or more current carrying capacity as the copper conductors. Also, the conduit sizes shall be increased accordingly.

(4) All conductors with the size of No. 8 or larger shall be stranded.

(5) All lighting and receptacle branch circuits in excess of 100 linear feet shall be increased one size to prevent excessive voltage drop.

4. SAFETY SWITCHES (FSS) (NFSS):

a. GENERAL: Furnish and install safety switches as indicated on the drawings or as required. All safety switches shall be NEMA General Duty Type and Underwriters' Laboratories Listed. The switches shall be Fused Safety Switches (FSS) or Non-fused Safety Switches (NFSS) as shown on the drawings or required.

b. SWITCHES: Switches shall have a quick-make and quick-break operating handle and mechanism which shall be an integral part of the box. Padlocking provisions shall be provided for padlocking in the "OFF" position with at least three padlocks. Switches shall be horsepower rated for 250 volts AC or DC as required. Lugs shall be UL Listed for copper and aluminum cable.

c. ENCLOSURES: Switches shall be furnished in NEMA 1 general purpose enclosures with knockouts unless otherwise noted or required. Switches located on the exterior of the building or in "wet" locations shall have NEMA 3R enclosures (WP).

d. INSTALLATION: The safety switches shall be securely mounted in accordance with the N.E.C. The Contractor shall provide all mounting materials. Install fuses in the FSS. The fuses shall be dual element on motor circuits.

e. MANUFACTURER: SQUARE "D", GENERAL ELECTRIC, CUTLER-HAMMER or WESTINGHOUSE, or ITE.

5. PANELBOARDS - CIRCUIT BREAKER:

a. GENERAL: Furnish and install circuit breaker panelboards as indicated in the panelboard schedule and where shown on the drawings. The panelboard shall be dead front safety type equipped with molded case circuit breakers and shall be the type as listed in the panelboard schedule: Service entrance panelboards shall include a full capacity box bonding strap and approved for service entrance. The acceptable manufacturer of the panelboards are "ITE" GENERAL ELECTRIC, CUTLER-HAMMER and WESTINGHOUSE provided they are fully equal to the type listed on the drawings. The panelboard shall be listed by Underwriters' Laboratories and bear the UL Label.

b. CIRCUIT BREAKERS: Provide molded case circuit breakers of frame, trip rating and interrupting capacity as shown on the schedule. Also, provide the number of spaces for future circuit breakers as shown in the schedule. The circuit breakers shall be quick-make, quick-break, thermal-magnetic, trip indicating and have common trip on all multipole breakers with internal tie mechanism.

c. PANELBOARD BUS ASSEMBLY: Bus bar connections to the branch circuit breakers shall be the "phase sequence" type. Single phase 3-wire panelboard bussing shall be such that any two adjacent single pole breakers are connected to opposite polarities in such a manner that 2-pole breakers can be installed in any location. Three phase 4-wire bussing shall be such that any three adjacent single pole breakers are individually connected to each of the three different phases in such a manner that 2 or 3-pole breakers can be installed at any location. All current carrying parts of the bus assembly shall be plated. Mains ratings shall be as shown in the panelboard schedule on the plans. Provide solid neutral assembly (S/N) when required.

d. WIRING TERMINALS: Terminals for feeder conductors to the panelboard mains and neutral shall be suitable for the type of conductor specified. Terminals for branch circuit wiring, both breaker and neutral, shall be suitable for the type of conductor specified.

e. CABINETS AND FRONTS: The panelboard bus assembly shall be enclosed in a steel cabinet. The size of the wiring gutters and gauge of steel shall be in accordance with NEMA Standards. The box shall be fabricated from galvanized steel or equivalent rust resistant steel. Fronts shall include door and have flush, brushed stainless steel, spring-loaded door pulls. The flush lock shall not protrude beyond the front of the door. All panelboard locks shall be keyed alike. Fronts shall not be removable with door in the locked position.

f. DIRECTORY: On the inside of the door of each cabinet, provide a typewritten directory which will indicate the location of the equipment or outlets supplied by each circuit. The directory shall be mounted in a metal frame with a non-breakable transparent cover. The panelboard designation shall be typed on the directory card and panel designation stenciled in 1-1/2" high letters on the inside of the door.

g. PANELBOARD INSTALLATION:

(1) Before installing panelboards check all of the architectural drawings for possible conflict of space and adjust the location of the panelboard to prevent such conflict with other items.

(2) When the panelboard is recessed into a wall serving an area with accessible ceiling space, provide and install an empty conduit system for future wiring. A 1-1/4" conduit shall be stubbed into the ceiling space above the panelboard and under the panelboard if such accessible ceiling space exists.

(3) The panelboards shall be mounted in accordance with Article 373 of the N.E.C. The Electrical Contractor shall furnish all material for mounting the panelboards.

6. WIRING DEVICES:

a. GENERAL: The wiring devices specified below with ARROW HART numbers may also be the equivalent wiring device as manufactured by BRYANT ELECTRIC, HARVEY HUBBELL or PASS & SEYMOUR. All other items shall be as specified.

b. WALL SWITCHES: Where more than one flush wall switch is indicated in the same location, the switches shall be mounted in gangs under a common plate.

(1)	Single Pole	AH#1991
(2)	Three-Way	AH#1993
(3)	Four-Way	AH#1994
(4)	Switch with pilot light	AH#2999-R
(5)	Motor Switch - Surface	AH#6808
(6)	Motor Switch - Flush	AH#6808-F

c. RECEPTACLES:

(1)	Duplex	AH#6739
(2)	Clock Outlet	AH#5708
(3)	Weatherproof	AH#5735-F

(4) Floor Receptacles - Steel City Series 600 floor box with bronze edge ring, floor plate P-60-1, bronze carpet plate and service fitting SFH-40-RG.

(5) Floor Outlet for Telephone and Alarm - Steel City Series 600 floor box with bronze edge ring, floor plate P-60-1, bronze carpet plate and service fitting SFL-10.

d. WALL PLATE: Stainless steel wall plates with satin finish miminum .030 inches shall be provided for all outlets and switches.

SECTION 16E - LIGHTING FIXTURES AND LAMPS

1. Portions of the sections of the documents designated by the letters "A", "B" & "C" and "DIVISION ONE - GENERAL REQUIREMENTS" apply to this Division.

2. Consult Index to be certain that set of Documents and Specifications is complete. Report omissions or discrepancies to the Architect.

3. LIGHTING FIXTURES:

    a. GENERAL: The Electrical Contractor shall furnish, install and connect all lighting fixtures to the building wiring system unless otherwise noted.

    b. FIXTURE TYPE: The fixture for each location is indicated by type letter. Refer to fixture schedule on the drawings for each type, manufacturer, catalog number and type of mounting.

    c. FLUORESCENT BALLASTS: All fluorescent fixtures shall have ETL-CBM high power factor, quiet operating, Class "A" sound rated, thermally protected Class "P" cool-rated ballast with UL approval. Ballasts shall be as manufactured by GENERAL ELECTRIC, ADVANCE, JEFFERSON or approved equal. The ballasts shall be subject to a two (2) year manufacturer's guarantee. Guarantee shall be filed with the Owner.

    d. SHOP DRAWINGS:

        (1) Shop drawings for lighting fixtures shall indicate each type together with manufacturer's name and catalog number, complete photometric data compiled by an independent testing laboratory and type of lamp (s) to be installed. No fixtures shall be delivered to the job until approved by the Architect.

        (2) If the Electrical Contractor submits shop drawings on a fixture for approval other than those specified, he shall also submit a sample fixture when requested by the Architect. The sample fixture will be returned to the Electrical Contractor. The decision of the Architect shall be final.

e. COORDINATION: It shall be the responsibility of the Electrical Contractor to coordinate with the ceiling contractor and the General Contractor in order that the proper type fixture be furnished to match the ceiling suspension system being installed or building construction material.

4. LAMPS:

a. The Electrical Contractor shall furnish and install lamps in all fixtures as indicated on the drawings or as required. Fluorescent lamps shall be standard cool white and incandescent lamps shall be inside frosted unless otherwise noted on the drawings.

b. Lamps shall be manufactured by General Electric, Westinghouse or Sylvania.

SECTION 16F - SPECIAL SYSTEMS

1. Portions of the sections of the Documents designated by the letters "A", "B" & "C" and "DIVISION ONE - GENERAL REQUIREMENTS" apply to this Division.

2. Consult Index to be certain that set of Documents and Specifications is complete. Report omissions or discrepancies to the Architect.

3. TELEPHONE RACEWAY SYSTEMS:

    a. GENERAL: The Electrical Contractor shall provide and install empty raceway, outlet boxes, pull boxes and associated equipment required for a complete telephone system as indicated on the drawings and specified herein. All materials and workmanship shall conform with Section 16B of the Specifications. All wiring shall be installed by the local telephone company. The entire installation shall be in accordance with the requirements of the local telephone company.

    b. RIGID CONDUIT (Heavy Wall): Rigid conduit shall be installed in the following locations: service entrance, underground in contact with earth, in concrete slab and "wet" locations.

    c. ELECTRIC METALLIC CONDUIT (EMT): Electric metallic tubing shall be used in all locations not otherwise specified to be rigid conduit.

    d. OUTLETS: Telephone wall outlets shall consist of a 4" two gang outlet box, raised device cover and a telephone device plate of the same material as the receptacle device plates. The conduit shall extend from the outlet to the designated telephone space unless otherwise noted.

    e. PULL WIRE: The Electrical Contractor shall install a No. 14 ga. galvanized pull wire in the raceway system for future use.

    f. MOUNTING HEIGHTS: The wall outlets shall be mounted at approximately the following heights unless otherwise noted on the drawings or required by telephone company: Desk Phones - 18" AFF, Wall Phone - 58" AFF, Telephone Booth - 7'-6" AFF.

4. <u>EMERGENCY LIGHTING SYSTEM</u>:

a. The Electrical Contractor shall provide and install a complete emergency lighting system as indicated on the drawings and specified herein. The system shall originate on the line side of the service entrance main switch, through overcurrent protective equipment to each exit light fixture and each fixture designated as being "emergency light". The switch shall be painted red. The Contractor shall be responsible for verification with local governing authorities of the proper letter and background colors of exit light fixtures before purchase of same. The entire installation shall be in accordance with the National Electric Code, the local electric code and the fire protection department having authority in the local jurisdiction.

ASSIGNMENT 9

Answer the following questions by filling in the blank spaces.

1. Section _____—General Provisions.

2. The paragraph in the specifications that gives information concerning inspection and certificates of approval is number _____, subparagraph _____, and is in Section _____—General _____.

3. In Section 16B—"Basic Materials and Workmanship"—conduit, couplings, and connectors shall be as manufactured by _____ or _____.

4. This same section also specifies that exposed conduit shall be run _____ to or at right angles with the _____.

5. Conduit throughout the project shall be securely and rigidly supported to the _____.

6. Single conduit, 1¼ inches and larger, that runs concealed horizontally shall be supported by suitable _____ _____.

7. Outlets boxes located on the exterior in damp or wet locations or as otherwise noted shall be threaded cast-aluminum device boxes such as _____ type _____ or _____.

8. Wall receptacles shall be mounted approximately _____ above the finished floor.

9. Wire shall be as manufactured by _____ _____ or approved equal.

10. All joints or splices for No. 10 AWG wire or smaller shall be made with UL approved _____ or _____ connectors.

11. Motor-protection fuses shall be _____ element.

12. No structural members shall be cut without the approval of the _____.

13. The electrical contractor shall provide all flashing, caulking, and sleeves required where his items pass through the outside _____ or _____.

14. The electrical contractor shall properly ground the electrical system as required by the _____.

15. The conduit used for the service entrance shall be _____ unless otherwise noted on the drawings.

16. The minimum size of conduit allowed on this project is _____.

17. The conductor material for feeders shall be type RHW or THW rated at _____°C.

18. No conductor shall be smaller than No. _____.

19. Circuit-breaker panelboards shall be furnished and installed as indicated in the panelboard _____ and where shown on the _____.

20. The catalog number for single-pole switches is AH _____.

21. The specifications refer the contractor to the _____ schedule on the drawings for the type, manufacturer, catalog number, and mounting of each lighting fixture.

22. The electrical contractor shall install a No. 14 gauge _____ in the telephone raceway system for future use.

10

Reproduction of Drawings

The original tracing of an electrical drawing is often very valuable because of the amount of work required to prepare it. If the original drawing was used by every person needing to refer to it, the drawing would soon become damaged, badly worn, and, in time, destroyed. Therefore, some inexpensive and rapid means of making exact copies is needed. These copies can then be distributed to those who need the information on the drawings.

The most popular means of reproducing original tracings of drawings is by the blueline method. This method gives a finished print of dark lines on a white background. Another method, blueprinting, is widely used also. It produces a print consisting of white lines on a blue background.

BLUEPRINTING

Blueprinting frames have been used for years in reproducing original tracings. A simple blueprinting form consisted of a flat surface, usually of wood, and covered with a padding of soft felt material. A glass frame was hinged to the form. The print was made on paper that had been coated with chemicals sensitive to sunlight. This blueprint paper was laid on the felt surface of the frame with its coated side up, the tracing was laid over it right side up, and then the glass was pressed down firmly and fastened in place.

When the drawing had remained in the sun for a few minutes, the blueprint paper was taken out and thoroughly washed in clean water for several minutes and then hung up to dry. If the exposure was timed correctly, the coated surface of the paper would now be a clear, dark blue color except where it was covered by the lines on the drawing—these were perfectly white.

While the method described is rather obsolete, it did offer an inexpensive method of reproducing drawings made on tracing paper.

A more modern method of obtaining blueprints (although at a much greater cost in equipment) is the electric blueprinting machine. This type of machine passes the original tracing, along with the coated paper, around a glass cylinder containing electric arc lamps. The speed at which the paper travels may be adjusted to suit the quality of the tracing, as far as the intensity and size of the lines and the depth of the background are concerned.

BLUELINE OR WHITEPRINTING

Another method of reproducing original drawings uses a machine commonly called a whiteprinter. This method consists of exposing chemically treated paper and the original tracing to a high-intensity discharge or fluorescent lamp. The ultraviolet light reduces the part of the sensitized surface that is unprotected by the lines of the original tracing into an invisible compound. After exposure, the print (only) is then subjected to ammonia vapors which develop the sensitized lines. The finished print is true to scale and ready for immediate use. Since this process is completely dry, it is the most popular for most applications. Depending on the type of paper that is used, the lines printed with this process may be red, brown, blue, or black on a white background.

PHOTOCOPYING

Another very convenient method of reproducing drawings on any material is to make photocopies or photostats with a photocopier. This method actually takes a picture of the original drawing and then produces as many prints as desired. This type of copier also has the advantage of making it possible to reduce the original to any size preferred.

Photocopying has many other advantages that can save hours of drafting time. For example, sometimes original drawings become creased, stained, or worn to the point that they cannot be satisfactorily reproduced, revised, or microfilmed. Of course, the original drawing can be redrawn or traced, but the expense is high. However, there are photographic materials and techniques that can be used to restore old worn drawings to new usefulness. Fig. 10-1 shows an example of a worn drawing. Fig. 10-2 shows this same drawing after it has been restored by photographic techniques.

Few drawings never need revisions or changes from time to time. Again, the photocopier can save valuable time. Rather than retrace a drawing that needs only a few changes, a photostat print can be made and the unwanted part cut out. After taping the remaining drawing to a new drawing form, the composite is photographed and printed on a special base film. The revisions can then be made on this film. Much time is saved by redrawing only those portions actually needing revision.

As mentioned previously, photocopiers can also be used to change the size of the original drawing. This is valuable when you want to reduce the size to save file space or cut printing and postage costs, or when you want to enlarge the drawing to open up detail clutter for greater legibility or to make revisions easier.

MICROFILMING

Microfilming is rapidly moving into the electrical industry as its convenience and space-saving features are recognized. For example, suppose a firm required 150 cubic feet to store all of its drawings. This storage space could be reduced to 1 cubic foot if the drawings were transferred to microfilm. Microfilming not only saves space but also saves postage (weight) if many drawings have to be mailed.

The first step in microfilming drawings is to convert the original drawing to a microfilm frame by means of a special camera mounted on a table. The camera reduces a drawing so that it will fit on the microfilm; the reduction can be varied by changing the height of the camera above the drawing.

After the film is exposed, it is passed through a processor, which develops it, and is then mounted on some type of card or frame.

Once the film has been mounted, the image can be blown up for viewing by means of a special viewer. Also

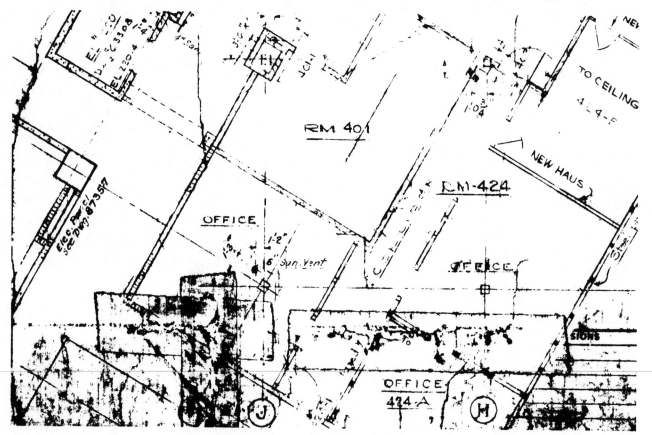

Fig. 10-1. Example of a worn drawing before restoration.

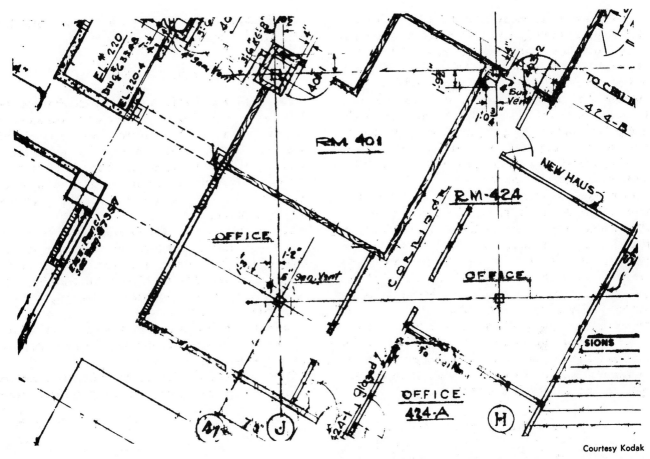

Courtesy Kodak

Fig. 10-2. Same drawing as shown in Fig. 10-1 after being restored by photographic techniques.

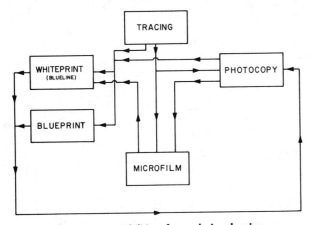

Fig. 10-3. Various possibilities of reproducing drawings.

available are viewer-printers that will make full-size prints from the microfilm.

With the great amount of reduction used in microfilming, high-quality drawings should be used; that is, everything on the drawing must be sharp and uncluttered. All numbers and lettering should have a minimum height, especially when a government contract is involved.

All of these processes are, obviously, more expensive than the blueprinting process. Fig. 10-3 is a flow chart showing various ways of reproducing drawings.

Glossary

Accessible (as applied to wiring methods)—Capable of being removed or exposed without damaging the building structure or finish, or not permanently closed in by the structure or finish of the building.

Aggregate—Inert material mixed with cement and water to produce concrete.

Ampacity—Current-carrying capacity expressed in amperes.

Backfill—Loose earth placed outside foundation walls for filling and grading.

Bearing plate—Steel plate placed under one end of a beam or truss for load distribution.

Bearing wall—Wall supporting a load other than its own weight.

Bench Mark—Point of reference from which measurements are made.

Bonding jumper—A reliable conductor used to ensure the required electrical conductivity between metal parts required to be electrically connected.

Branch circuit—That portion of the wiring system between the final overcurrent device protecting the circuit and the outlet(s).

Bridging—System of bracing between floor beams to distribute floor load.

Building—A structure that stands alone or that is cut off from adjoining structures by fire walls with all openings therein protected by approved fire doors.

Cavity wall—Wall built of solid masonry units arranged to provide air space within the wall.

Chase—Recess in inner face of masonry wall providing space for pipes and/or ducts.

Column—Vertical load-carrying member of a structural frame.

Concealed—Rendered inaccessible by the structure or finish of the building. Wires in concealed raceways are considered concealed even though they may become accessible by withdrawing them.

Conductor—A *bare* conductor is one having no covering or insulation whatsoever. A *covered* conductor is one having one or more layers of nonconducting materials that are not recognized as insulation under the National Electrical Code. An *insulated* conductor is one covered with material recognized as insulation.

Connector, pressure (solderless)—A connector that establishes the connection between two or more conductors, or between one or more conductors and a terminal by means of mechanical pressure and without the use of solder.

Continuous load—A load in which the maximum current is expected to continue for three hours or more.

Contour line—On a land map denoting elevations, a line connecting points with the same elevation.

Crawl space—Shallow space between the first tier of beams and the ground (no basement).

Curtain wall—Nonbearing wall between piers or columns for the enclosure of the structure; not supported at each story.

Demand factor—In any system or part of a system, the ratio of the maximum demand of the system, or part of the system, to the total connected load of the system, or part of the system under consideration.

Disconnecting means—A device, a group of devices, or other means with which the conductors of a circuit can be disconnected from their source of supply.

Double-strength glass—Sheet glass that is one-eighth inch thick (single-strength glass is one-tenth inch thick).

Dry wall—Interior wall construction consisting of plaster boards, wood paneling, or plywood nailed directly to the studs without application of plaster.

Elevation—Drawing showing the projection of a building on a vertical plane.

Expansion bolt—Bolt with a casing arranged to wedge the bolt into a masonry wall to provide an anchorage.

Expansion joint—Joint between two adjoining concrete members arranged to permit expansion and contraction with changes in temperature.

Exposed (as applied to live parts)—Live parts that a person could inadvertently touch or approach nearer than a safe distance. This term is applied to parts not suitably guarded, isolated, or insulated.

Exposed (as applied to wiring methods)—On or attached to the surface or behind panels to allow access.

Facade—Main front of a building.

Feeder—The conductors between the service equipment, or the generator switchboard of an isolated plant, and the branch-circuit overcurrent device.

Fire stop—Incombustible filler material used to block interior draft spaces.

Flashing—Strips of sheet metal bent into an angle between the roof and wall to make a watertight joint.

Footing—Structural unit used to distribute loads to the bearing materials.

Frost line—Deepest level below grade to which frost penetrates in a geographic area.

Ground—A conducting connection, whether intentional or accidental, between an electrical circuit or piece of equipment, and earth or some other conducting body serving in place of the earth.

Grounded—Connected to earth or to some conducting body that serves in place of the earth.

Grounded conductor—A system or circuit conductor that is intentionally grounded.

Grounding conductor—A conductor used to connect equipment or the grounded circuit of a wiring system to a grounding electrode or electrodes.

I beam—Rolled steel beam or built-up beam of I section.

Incombustible material—Material that will not ignite or actively support combustion in a surrounding temperature of 1200°F during an exposure of 5 minutes; also, material that will not melt when the temperature of the material is maintained at 900°F for a period of at least 5 minutes.

Jamb—Upright member forming the side of a door or window opening.

Lally column—Compression member consisting of a steel pipe filled with concrete under pressure.

Laminated wood—Wood built up of plies or laminations that have been joined either with glue or with mechanical fasteners. Usually, the plies are too thick to be classified as veneer and the grain of all plies is parallel.

Lighting outlet—An outlet intended for the direct connection of a lampholder, a lighting fixture, or a pendant cord terminating in a lampholder.

Nonbearing wall—Wall that carries no load other than its own weight.

Outlet—A point on the wiring system at which current is taken to supply utilization equipment.

Panelboard—A single panel or group of panel units designed for assembly in the form of a single panel, including buses, and with or without switches and/or automatic overcurrent protective devices for the control of light, heat, or power circuits of small individual as well as aggregate capacity; designed to be placed in a cabinet or cutout box placed in or against a wall or partition and accessible only from the front.

Pilaster—Flat square column attached to a wall and projecting about a fifth of its width from the face of the wall.

Plenum—Chamber or space forming a part of an air-conditioning system.

Precast concrete—Concrete units (such as piles or vaults) cast away from the construction site and set in place.

Raceway—Any channel designed expressly for holding wires, cables, or bus bars and used solely for this purpose.

Rainproof—So constructed, protected, or treated as to prevent rain from interfering with successful operation of the apparatus.

Raintight—So constructed or protected that exposure to a beating rain will not result in the entrance of water.

Readily accessible—Capable of being reached quickly, for operation, renewal, or inspections, without requiring those to whom ready access is requisite to climb over or remove obstacles or resort to portable ladders, chairs, etc.

Receptacle—A contact device installed at the outlet for the connection of a single attachment plug.

Riser—Upright member of stair extending from tread to tread.

Roughing in—Installation of all concealed electrical wiring; includes all electrical work done before finishing.

Service—The conductors and equipment used for delivering energy from the electricity supply system to the wiring system of the premises being served.

Service cable—The service conductors made up in the form of a cable or wiring assembly.

Service conductors—The supply conductors that extend from the street main or from transformers to the service equipment of the premises being supplied.

Service drop—The overhead service conductors from the last pole or other aerial support to and including the splices, if any, connecting to the service-entrance conductors at the building or other structure.

Service-entrance conductors, overhead system—The service conductors between the terminals of the service equipment and a point usually outside the building, clear of building walls, where joined by tap or splice to the service drop.

Service-entrance conductors, underground system—The service conductors between the terminals of the service equipment and the point of connection to the service lateral.

Service equipment—The necessary equipment, usually consisting of a circuit breaker or a switch and fuses, and their accessories, located near the point of entrance of supply conductors to a building or other structure, or an otherwise defined area, and intended to constitute the main control and means of cutoff of the supply.

Service lateral—The underground service conductors between the street main, including any risers at a pole, other structure, or from transformers, and the first point of connection to the service-entrance conductors in a terminal box, meter, or other enclosure with adequate space, inside or outside the building wall. Where there is no terminal box, meter, or other enclosure with adequate space, the point of connection shall be considered to be the point of entrance of the service conductors into the building.

Service raceway—The rigid metal conduit, electrical metallic tubing, or other raceway, that encloses the service-entrance conductors.

Sheathing—First covering of boards or paneling nailed to the outside of the wood studs of a frame building.

Siding—Finishing material that is nailed to the sheathing of a wood frame building and that forms the exposed surface.

Signal circuit—Any electrical circuit that supplies energy to an appliance that gives a recognizable signal.

Soffit—Underside of a stair, arch, or cornice.

Soleplate—Horizontal bottom member of wood-stud partition.

Studs—Vertically set skeleton members of a partition or wall to which lath is nailed.

Switch, general-use—A switch intended for use in general distribution and branch circuits. It is rated in amperes and is capable of interrupting its rated current at its rated voltage.

Also, a form of switch so constructed that it can be installed in flush device boxes or on outlet-box covers, or otherwise used in conjunction with wiring systems recognized by the National Electrical Code.

Switch, general-use ac snap—A form of switch suitable only for use on alternating-current circuits for controlling the following:

1. Resistive and inductive loads (including electric-discharge lamps) not exceeding the ampere rating at the voltage involved.
2. Tungsten-filament lamp loads not exceeding the ampere rating at 120 volts.
3. Motor loads not exceeding 80% of the ampere rating of the switches at the rated voltage.

All ac general-use snap switches are marked *ac* in addition to their electrical rating.

Switch, general-use, ac-dc snap—A form of switch suitable for use on either direct- or alternating-current circuits for controlling the following:

1. Resistive loads not exceeding the ampere rating at the voltage involved.
2. Inductive loads not exceeding 50% of the ampere rating at the voltage involved, except that switches having a marked horsepower rating are suitable for controlling motors not exceeding the horsepower rating of the switch at the voltage involved.
3. Tungsten-filament lamp loads not exceeding the ampere rating at 125 volts, when marked with the letter *T*.

Ac-dc general-use snap switches generally are not marked *ac-dc,* but are always marked with their electrical rating.

Switch, isolating—A switch intended for isolating an electric circuit from the source of power. It has no interrupting rating and is intended to be operated only after the circuit has been opened by some other means.

Switchboard—A large single panel, frame, or assembly of panels, having switches, overcurrent and other protective devices, buses, and usually instruments, mounted on the face or back or both. Switchboards are generally accessible from the rear as well as from the front and are not intended to be installed in cabinets.

Trusses—Framed structural pieces consisting of triangles in a single plane for supporting loads over spans.

Voltage—The greatest root-mean-square (effective) difference of potential between any two conductors of the circuit concerned.

Voltage to ground—In grounded circuits, the voltage between the given conductor and the point or conductor of the circuit that is grounded; in ungrounded circuits, the greatest voltage between the given conductor and any other conductor of the circuit.

Watertight—So constructed that moisture will not enter the enclosing case.

Weatherproof—So constructed or protected that exposure to the weather will not interfere with successful operation.

Web—Central portion of an I beam.

Final Assignment

The set of electrical drawings found in this section is typical of those used in building construction and will be used for the final examination. The reader should review Chapter 3, "Electrical Symbols," prior to taking the exam.

TRUE-FALSE QUESTIONS

Check the correct answer for the following true-false questions. If any part of the statement is false, the statement should be marked false.

	TRUE	FALSE
1. The type-6 lighting fixtures shown on the floor plan are fluorescent lighting fixtures (Fig. B-1).	_____	_____
2. The motor starter used for the compressor (comp.) motor utilizes a thermal overload relay (Fig. B-2).	_____	_____
3. Panel A is shown on the floor plans (Figs. B-1 and B-2) and also in the power-riser diagram (Fig. B-3).	_____	_____
4. A description of panel A, that is, the size of mains, breakers, etc., is found in the power-riser diagram (Fig. B-3).	_____	_____
5. The safety switch shown on the power floor plan (Fig. B-2) and fed by circuit No. A-33 contains 20-ampere fuses.	_____	_____
6. A slash mark through a conventional duplex receptacle symbol means that it is mounted 18 inches above the finished floor.	_____	_____
7. The six Sq. "D" disconnects—Cat. No. HU221 (Fig. B-2)—are fusible types.	_____	_____
8. There are ten type-5 lighting fixtures shown on the lighting floor plan (Fig. B-1).	_____	_____

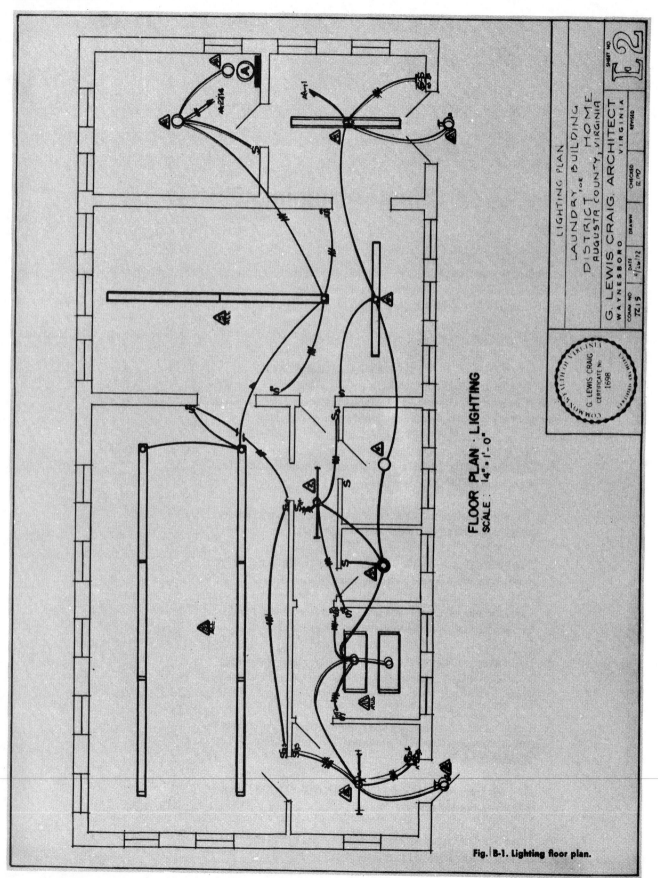

FLOOR PLAN · LIGHTING
SCALE: 1/4" = 1'-0"

LIGHTING PLAN
LAUNDRY for BUILDING
DISTRICT HOME
AUGUSTA COUNTY, VIRGINIA

G. LEWIS CRAIG. ARCHITECT
WAYNESBORO   VIRGINIA

COMM. NO	DATE	DRAWN	CHECKED	REVISED
7215	4/26/72		12-172	

SHEET NO
E2

Fig. B-1. Lighting floor plan.

9. A total of three branch circuits are used to feed all of the lighting fixtures on this project (Fig. B-1). _____ _____

10. Fixture type 2 (Fig. B-1) is a bare tube fluorescent strip, as can be seen from the symbol list in Chapter 3. _____ _____

MULTIPLE CHOICE

Check one answer only for each of the following questions.

11. Type-1 lighting fixtures (Fig. B-3) contain:
    A. One 40-watt fluorescent lamp. _____
    B. Three Par-38 flood lamps. _____
    C. Four 40-watt fluorescent lamps. _____
    D. One 60-watt incandescent lamp. _____

12. The circuit feeding the ironer in the building (Fig. B-2):
    A. Terminates in a junction box directly from the panelboard. _____
    B. Terminates in a junction box from a nonfusible disconnect which, in turn, is fed directly from panel A. _____
    C. Consists of three conductors. _____
    D. Is connected to circuit No. A-15. _____

13. Circuit No. "A-17" (Fig. B-2) is provided for:
    A. A drinking fountain. _____
    B. A power roof ventilator. _____
    C. A future washer. _____
    D. A water softener. _____

14. Panel A (Fig. B-3) contains only:
    A. Twelve 15-A circuit breakers. _____
    B. Seven 20-A circuit breakers. _____
    C. Seven 30-A circuit breakers. _____
    D. Twelve "provisions only." _____

15. The Lithonia lighting fixture (Fig. B-3)—Cat. No. C240—contains:
    A. The same type of lamp as fixture type 1. _____
    B. The same number of lamps as fixture type 1. _____
    C. One 60-watt fluorescent lamp. _____
    D. Four 40-watt flourescent lamps. _____

FILL INS

Answer the following questions by filling in the blanks.

16. The 2-P main circuit breakers (Fig. B-3) in the existing main electric panel, A, are rated at _____ amperes.

17. The feeder circuit from the existing panel to new panel A (Fig. B-3) consists of three _____ conductors in _____ conduit.

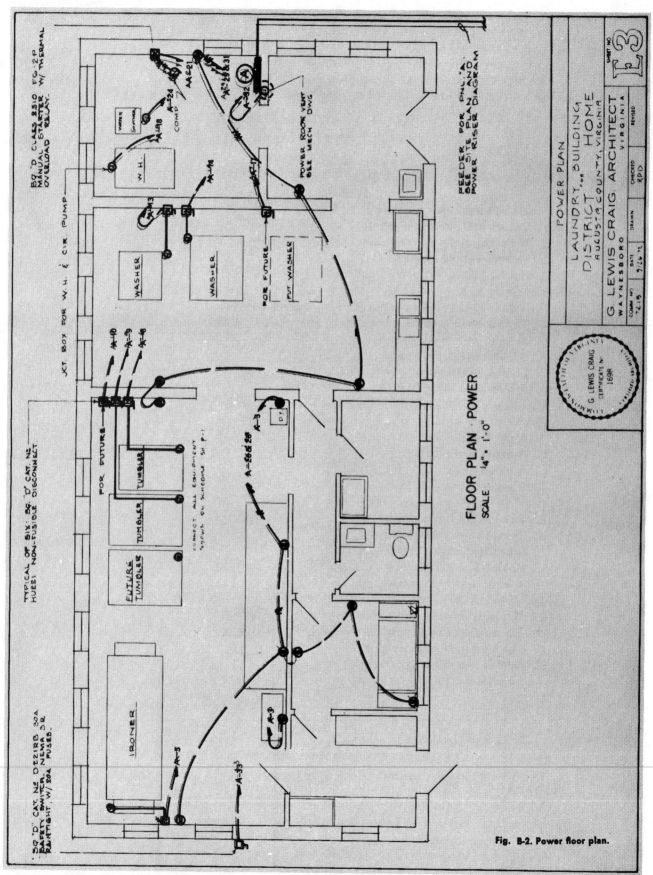

FLOOR PLAN · POWER
SCALE 1/4" = 1'-0"

Fig. B-2. Power floor plan.

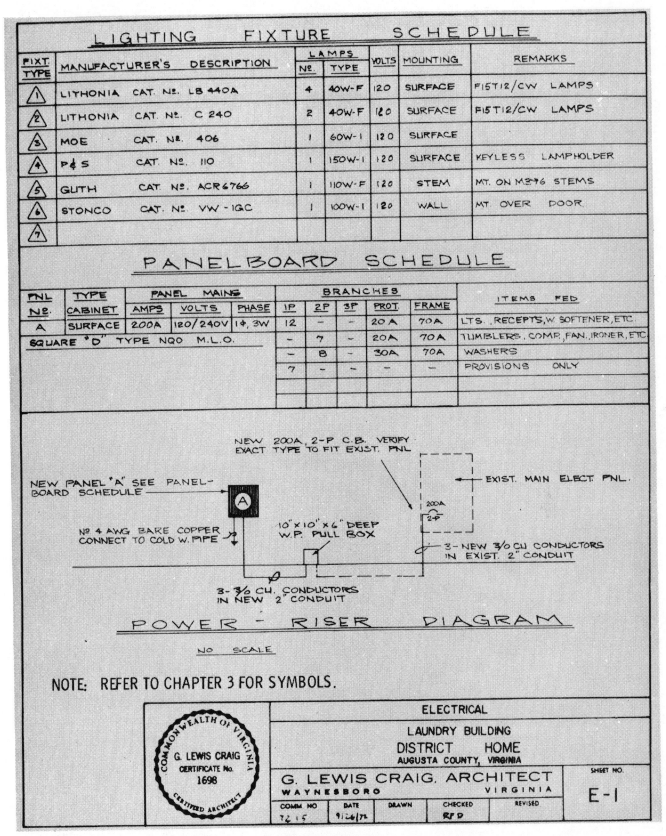

LIGHTING FIXTURE SCHEDULE

FIXT. TYPE	MANUFACTURER'S DESCRIPTION	LAMPS No.	LAMPS TYPE	VOLTS	MOUNTING	REMARKS
1	LITHONIA CAT. No. LB 440A	4	40W-F	120	SURFACE	F15T12/CW LAMPS
2	LITHONIA CAT. No. C 240	2	40W-F	120	SURFACE	F15T12/CW LAMPS
3	MOE CAT. No. 406	1	60W-1	120	SURFACE	
4	P & S CAT. No. 110	1	150W-1	120	SURFACE	KEYLESS LAMPHOLDER
5	GUTH CAT. No. ACR6766	1	110W-F	120	STEM	MT. ON M376 STEMS
6	STONCO CAT. No. VW-1GC	1	100W-1	120	WALL	MT. OVER DOOR.
7						

PANELBOARD SCHEDULE

PNL No.	TYPE CABINET	PANEL MAINS AMPS	PANEL MAINS VOLTS	PANEL MAINS PHASE	BRANCHES 1P	BRANCHES 2P	BRANCHES 3P	BRANCHES PROT.	BRANCHES FRAME	ITEMS FED
A	SURFACE	200A	120/240V	1∅, 3W	12	–	–	20A	70A	LTS., RECEPTS, W. SOFTENER, ETC.
SQUARE "D" TYPE NQO M.L.O.					–	7	–	20A	70A	TUMBLERS, COMP, FAN, IRONER, ETC
					–	8	–	30A	70A	WASHERS
					7	–	–	–	–	PROVISIONS ONLY

NEW 200A, 2-P C.B. VERIFY EXACT TYPE TO FIT EXIST. PNL

NEW PANEL 'A' SEE PANEL-BOARD SCHEDULE → A

← EXIST. MAIN ELECT. PNL.

200A 2-P

No. 4 AWG BARE COPPER CONNECT TO COLD W. PIPE

10"X10"X6" DEEP W.P. PULL BOX

3-NEW 3/0 CU CONDUCTORS IN EXIST. 2" CONDUIT

3-3/0 CU. CONDUCTORS IN NEW 2" CONDUIT

POWER - RISER DIAGRAM

NO SCALE

NOTE: REFER TO CHAPTER 3 FOR SYMBOLS.

COMMONWEALTH OF VIRGINIA
G. LEWIS CRAIG
CERTIFICATE No.
1698
CERTIFIED ARCHITECT

ELECTRICAL

LAUNDRY BUILDING
DISTRICT HOME
AUGUSTA COUNTY, VIRGINIA

G. LEWIS CRAIG, ARCHITECT
WAYNESBORO   VIRGINIA

COMM. NO	DATE	DRAWN	CHECKED	REVISED
7215	9/26/72		RFD	

SHEET NO.
E-1

Fig. B-3. Power-riser diagram.

18. The type 5 lighting fixture (Fig. B-3) is manufactured by _____ and the catalog number is _____.

19. The type 6 lighting fixture (Fig B-1) is shown mounted on the _____.

20. Panel A is rated for 120/240 volts and the cabinet type (Fig. B-3) is designed for _____ mounting.

21. The grounding conductor for the electrical service (Fig. B-3) is _____ base copper and is connected to a _____ pipe.

22. The SQ. "D" Cat. No. D221RB safety switch (Fig. B-2) is an NEMA 3R which means that the switch is _____.

23. Panel A is shown on the power floor plan (Fig. B-2); further details on this panel may be found in the _____ diagram and the _____ schedule.

24. Both of the floor plans (Figs. B-1 and B-2) are drawn to a scale of _____.

25. The power-riser diagram (Fig. B-3) is drawn to _____ scale.

Answers to Assignments

ASSIGNMENT 1

1. An electrical blueprint is an exact copy or reproduction of an original drawing consisting of lines, symbols, dimensions, and notations to accurately convey an engineer's design to workmen who install the electrical system on the job.
3. Electrical construction
3. A. Draw to some given scale.
   B. Give dimensions on the drawing.
4. Electrical diagrams.
5. A. A plot plan.
   B. Floor plans.
   C. Power-riser diagram.
   D. Control-wiring schematic diagrams.
   E. Schedules.
   F. Notes.
   G. Large-scale details.
6. A. Electrical construction drawings.
   B. Single-line block diagrams.
   C. Schematic wiring diagrams.

ASSIGNMENT 2

1. A. 6′ 4″
   B. 12′ 8″
   C. 2′ 6″
   D. 4′ 9″
   E. 0′ 6″
2. K. 38′ 6″
   L. 16′ 0″
   M. 2′ 6″
   N. 2′ 9″
   F. 1′ 0″
   G. 2′ 0″
   H. 0′ 4″
   I. 1′ 9″
   J. 2′ 11″
   O. 7′ 0″
   P. 5′ 0″
   Q. 6′ 0″
   R. 7′ 0″

ASSIGNMENT 3

1. G
2. D
3. H
4. R
5. N
6. O
7. I
8. A
9. C
10. S
11. T
12. M
13. L
14. J
15. K
16. F
17. E
18. B
19. P
20. Q

ASSIGNMENT 4

1. 1500 watts
2. 5
3. junction box, 6
4. note, ¾, the floor
5. E
6. Recessed
7. Moldcast, A-270
9. surface
10. 2

ASSIGNMENT 5

1. sliced, sawed
2. clarify
3. 8 inches
4. sleeve, chase

5. visualization
6. detail
7. A 4-inch-square box
8. 1 inch
9. 1¼ inch
10. 70 kilowatts

ASSIGNMENT 6

1. three, one
2. T_1, T_2, T_3
3. 4160 volts
4. Power
5. 2000 amperes
6. ¾ inch, 1 inch
7. water-pipe
8. 14, ½
9. First floor
10. National Electrical Code

ASSIGNMENT 7

1. 18.3
2. 8
3. 4
4. 600
5. 1200
6. A
7. ⅓
8. 72 inches above finished floor

9. 1½ inch
10. 250

ASSIGNMENT 8

1. 13
2. 2
3. parking
4. A. 2300
   B. 3000
   C. 2300
   D. 3000
   E. 9750
5. 35 feet

ASSIGNMENT 9

1. 16A
2. 11, a, 16A, General Provisions
3. Triangle Conduit and Cable, or National Electric
4. parallel, lines of the building
5. building structure
6. beam clamps
7. Crouse Hinds FS, FD
8. 18 inches
9. General Cable, Triangle
10. wire nuts, compression-type
11. dual
12. Architect
13. walls, roof

Index

Other Practical References

Building Layout
Shows how to use a transit to locate the building on the lot correctly, plan proper grades with minimum excavation, find utility lines and easements, establish correct elevations, lay out accurate foundations and set correct floor heights. Explains planning sewer connections, leveling a foundation out of level, using a story pole and batterboards, working on steep sites, and minimizing excavation costs. **240 pages, 5½ x 8½, $11.75**

National Construction Estimator
Current building costs in dollars and cents for residential, commercial and industrial construction. Prices for every commonly used building material, and the proper labor cost associated with installation of the material. Everything figured out to give you the "in place" cost in seconds. Many time-saving rules of thumb, waste and coverage factors and estimating tables are included. **512 pages, 8½ x 11, $18.50. Revised annually.**

Blueprint Reading for the Building Trades
How to read and understand construction documents, blueprints, and schedules. Includes layouts of structural, mechanical and electrical drawings, how to interpret sectional views, how to follow diagrams; plumbing, HVAC and schematics, and common problems experienced in interpreting construction specifications. This book is your course for understanding and following construction documents. **192 pages, 5½ x 8½, $11.25**

Rough Carpentry
All rough carpentry is covered in detail: sills, girders, columns, joists, sheathing, ceiling, roof and wall framing, roof trusses, dormers, bay windows, furring and grounds, stairs and insulation. Many of the 24 chapters explain practical code approved methods for saving lumber and time without sacrificing quality. Chapters on columns, headers, rafters, joists and girders show how to use simple engineering principles to select the right lumber dimension for whatever species and grade you are using. **288 pages, 8½ x 11, $14.50**

Building Cost Manual
Square foot costs for residential, commercial, industrial, and farm buildings. In a few minutes you work up a reliable budget estimate based on the actual materials and design features, area, shape, wall height, number of floors and support requirements. Most important, you include all the important variables that can make any building unique from a cost standpoint. **240 pages, 8½ x 11, $14.00. Revised annually**

Concrete Construction & Estimating
Explains how to estimate the quantity of labor and materials needed, plan the job, erect fiberglass, steel, or prefabricated forms, install shores and scaffolding, handle the concrete into place, set joints, finish and cure the concrete. Every builder who works with concrete should have the reference data, cost estimates, and examples in this practical reference. **571 pages, 5½ x 8½, $20.50**

Estimating Electrical Construction
A practical approach to estimating materials and labor for residential and commercial electrical construction. Written by the A.S.P.E. National Estimator of the Year, it explains how to use labor units, the plan take-off and the bid summary to establish an accurate estimate. Covers dealing with suppliers, pricing sheets, and how to modify labor units. Provides extensive labor unit tables, and blank forms for use in estimating your next electrical job. **272 pages, 8½ x 11, $19.00**

Spec Builder's Guide
Explains how to plan and build a home, control your construction costs, and then sell the house at a price that earns a decent return on the time and money you've invested. Includes professional tips to ensure success as a spec builder: how government statistics help you judge the housing market, cutting costs at every opportunity without sacrificing quality, and taking advantage of construction cycles. Every chapter includes checklists, diagrams, charts, figures, and estimating tables. **448 pages, 8½ x 11, $24.00**

Residential Electrical Design
Explains what every builder needs to know about designing electrical systems for residential construction. Shows how to draw up an electrical plan from the blueprints, including the service entrance, grounding, lighting requirements for kitchen, bedroom and bath and how to lay them out. Explains how to plan electrical heating systems and what equipment you'll need, how to plan outdoor lighting, and much more. If you are a builder who ever has to plan an electrical system, you should have this book. **194 pages, 8½ x 11, $11.50**

Basic Plumbing with Illustrations
The journeyman's and apprentice's guide to installing plumbing, piping and fixtures in residential and light commercial buildings: how to select the right materials, lay out the job and do professional quality plumbing work. Explains the use of essential tools and materials, how to make repairs, maintain plumbing systems, install fixtures and add to existing systems. **320 pages, 8½ x 11, $17.50**

Electrical Construction Estimator
If you estimate electrical jobs, this is your guide to current material costs, reliable manhour estimates per unit, and the total installed cost for all common electrical work: conduit, wire, boxes, fixtures, switches, outlets, loadcenters, panelboards, raceway, duct, signal systems, and more. Explains what every estimator should know before estimating each part of an electrical system. **400 pages, 8½ x 11, $25.00**

Manual of Professional Remodeling
This is the practical manual of professional remodeling written by an experienced and successful remodeling contractor. Shows how to evaluate a job and avoid 30-minute jobs that take all day, what to fix and what to leave alone, and what to watch for in dealing with subcontractors. Includes chapters on calculating space requirements, repairing structural defects, remodeling kitchens, baths, walls and ceilings, doors and windows, floors, roofs, installing fireplaces and chimneys (including built-ins), skylights, and exterior siding. Includes blank forms, checklists, sample contracts, and proposals you can copy and use. **400 pages, 8½ x 11, $18.75**

Manual of Electrical Contracting
From the tools you need for installing electrical work in new construction and remodeling work to developing the finances you need to run your business. Shows how to draw up an electrical plan and design the correct lighting within the budget you have to work with. How to calculate service and feeder loads, service entrance capacity, demand factors, and install wiring in residential, commercial, and agricultural buildings. Covers how to make sure your business will succeed before you start it, and how to keep it running profitably. **224 pages, 8½ x 11, $17.00**